Ben Stacy Jerrik (Ed.)

Dinhata Subdivision

Ben Stacy Jerrik (Ed.)

Dinhata Subdivision

Cooch Behar district, West Bengal, Gram panchayat, Community development block in India

Part Press

Contents

Dinhata_subdivision

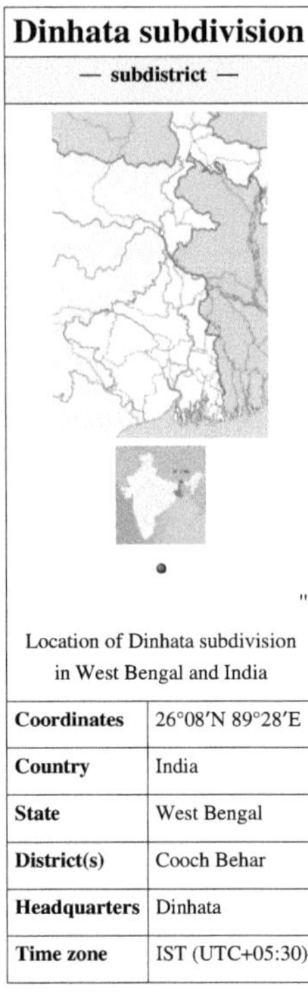

Dinhata subdivision

— subdistrict —

Location of Dinhata subdivision
in West Bengal and India

Coordinates	26°08′N 89°28′E
Country	India
State	West Bengal
District(s)	Cooch Behar
Headquarters	Dinhata
Time zone	IST (UTC+05:30)

Dinhata subdivision (Bengali: দিনহাটা মহকুমা) is a subdivision of the Cooch Behar district in the state of West Bengal, India. It consists of Dinhata municipality and three community development blocks: Dinhata–I, Dinhata–II and Sitai. The three blocks contain 33 gram panchayats and one census town. The subdivision has its headquarters at Dinhata.

Area

Apart from the Dinhata municipality, the subdivision contains one census town and rural areas of 33 gram panchayats under three community development blocks: Dinhata–I, Dinhata–II and Sitai.[1] The only census town under this block is Bhangri Pratham Khanda.[2]

Blocks

Dinhata–I block

Rural area under Dinhata–I block consists of 16 gram panchayats, viz. Bara Atiabari–I, Dinhata Village–II, Gosanimari–II, Putimari–I, Bara Atiabari–II, Gitaldaha–I, Matalhat, Putimari–II, Bara Soulmari, Gitaldaha–II, Okrabari, BHETAGURI–I, Dinhata Village–I, Gosanimari–I, Petla and Bhetaguri–II.[1] Urban area under this block comprises one census town: Bhangri Pratham Khanda.[2] Dinhata police station serves this block.[3] Headquarters of this block is in Dinhata.[4]

Dinhata–II block

Rural area under Dinhata–II block consists of 12 gram panchayats, viz. Bamanhat–I, Burirhat–II, Kishamat Dasgram, Sukarukuthi, Bamanhat–II, Chowdhurihat, Najirhat–I, Barasakdal, Gobra Chhara Nayarhat, Najirhat–II, Burirhat–I and Sahebganj.[1] There is no urban area under this block.[2] Dinhata police station serves this block.[3] Headquarters of this block is in Sahebganj.[4]

Sitai block

Rural area under Sitai block consists of five gram panchayats, viz. Adabari, Chamta, Sitai–II, Brahmottar–Chatra and Sitai–I.[1] There is no urban area under this block.[2] Sitai police station serves this block.[3] Headquarters of this block is in Sitai.[4]

Legislative segments

As per order of the Delimitation Commission in respect of the delimitation of constituencies in the West Bengal, the Dinhata municipality, Dinhata–II block and Bhetaguri–I, Dinhata Village–I, Dinhata Village–II and Putimari–I gram panchayats of Dinhata–I block together will constitute the Dinhata assembly constituency of West Bengal. The other twelve gram panchayats of Dinhata–I block, viz. Bara Atiabari–I, Gosanimari–II, Bara Atiabari–II, Gitaldaha–I, Matalhat, Putimari–II, Bara Soulmari, Gitaldaha–II, Okrabari, Gosanimari–I, Petla and Bhetaguri–II will form the Sitai assembly constituency along with the whole area under Sitai block. Sitai constituency will be reserved for Scheduled castes (SC) candidates. Both constituencies will be part of Cooch Behar (Lok Sabha constituency), which will be reserved for SC candidates.[5]

References

[1] "Directory of District, Sub division, Panchayat Samiti/ Block and Gram Panchayats in West Bengal, March 2008" (http://wbdemo5.nic.in/writereaddata/Directoryof_District_Block_GPs(RevisedMarch-2008).doc). *West Bengal*. National Informatics Centre, India. 2008-03-19. . Retrieved 2008-12-18.

[2] "District Wise List of Statutory Towns(Municipal Corporation,Municipality,Notified Area and Cantonment Board) , Census Towns and Outgrowths, West Bengal, 2001" (http://web.cmc.net.in/wbcensus/DataTables/01/Table-3.htm). Census of India, Directorate of Census Operations, West Bengal. . Retrieved 2008-12-18.

[3] "List of Districts/C.D.Blocks/ Police Stations with Code No., Number of G.Ps and Number of Mouzas" (http://web.cmc.net.in/wbcensus/DataTables/01/FrameTable2_1.htm). Census of India, Directorate of Census Operations, West Bengal. . Retrieved 2008-12-18.

[4] "Contact details of Block Development Officers" (http://wbdemo5.nic.in/html/asp/bdo_contact.asp?cd=DL). *Cooch Behar district*. Panchayats and Rural Development Department, Government of West Bengal. . Retrieved 2008-12-26.

[5] "Press Note, Delimitation Commission" (http://www.wbgov.com/e-gov/English/DELIMITATION.pdf) (PDF). *Assembly Constituencies in West Bengal*. Delimitation Commission. pp. 4,23. . Retrieved 2009-01-08.

Cooch_Behar_district

Cooch Behar district	
Location of Cooch Behar district in West Bengal	
State	West Bengal, India
Administrative division	Jalpaiguri
Headquarters	Cooch Behar
Area	3387 km^2 (**unknown operator: u'strong'** sq mi)
Population	2,822,780 (2011)
Population density	833 /km^2 (**unknown operator: u'strong'** /sq mi)
Urban population	225,618
Literacy	75.49 per cent[1]
Sex ratio	942
Lok Sabha constituencies	Cooch Behar
Assembly seats	Mekliganj, Mathabhanga, Cooch Behar Uttar, Cooch Behar Dakshin, Sitalkuchi, Sitai, Dinhata, Natabari, Tufanganj
Major highways	NH 31
Average annual precipitation	3201 mm
Official website [2]	

Cooch Behar district (Bengali: কোচবিহার জেলা, Rajbongshi/Kamatapuri : কোচবিহার) is a district of the state of West Bengal, India, as well as the district's namesake town. During the British Raj, the town of Cooch Behar was the seat of a princely state of Koch Bihar, ruled by the Koch dynasty.

As of 2011 it is the third least populous district of West Bengal (out of 19), after Dakshin Dinajpur and Darjeeling.[3]

Origin of name

The name "Cooch-Behar" is derived from the name of the Koch Rajbongshi tribe that is indigenous to this area. The word "Behar" is the Sanskrit word "Bihar" (to travel) which means the land through which the "Koch Rajbongshi" Kings used to travel or roam about ("Bihar").

The greatest Koch Rajbongshi King that has ever ruled in the Kingdom of Kamatapur is Maharaj Naranarayan, as well as his younger brother Prince Chilaray and other descendents. Historic Kamatapur comprises the total North Bengal maximum parts of Assam, some parts of present Bangladesh, Kishanganj district of Bihar and a few parts of Bhutan. The Koch-Rajbongshi community is demanding a separate state of their own comprising the parts of their old Kingdom to save their centuries-old culture from extinction.

History

The Koch dynasty originated from Mahishya community and has ruled the area around the town of Cooch Behar since the 16th century. The state remained unaffected by the great changes that overtook its surrounding provinces in the decade following the Battle of Plassey in 1757. However, it was invaded by Bhutan in the latter half of the 18th century, which prompted a British Ambassador to Bhutan, George Bogle to enter into a formal treaty alliance with the British in 1775. In 1947, the state acceded to the dominion of India and merged with the Union of India shortly afterwards.

Over time, Cooch Behar has been transformed from a kingdom to a state and from a state to the present status of a district. Before 28 August 1949, Cooch Behar was a Princely state ruled by the king of Cooch Behar, who had been a feudatory ruler under the British Government. By an agreement dated 28 August 1949 the king of Cooch Behar ceded full and extensive authority, jurisdiction and power of the state to the Dominion Government of India. The transfer of administration of the state to the Government of India came into force on 12 September 1949. Eventually, Cooch Behar was transferred and merged with the province of West Bengal on 19 January 1950 and from that date Cooch Behar emerged as a new District in the administrative map of West Bengal.

Geography

Cooch Behar is a district under the Jalpaiguri Division of the state of West Bengal. Cooch Behar is located in the northeastern part of the state and bounded by the district of Jalpaiguri in the north, state of Assam in the east and by Bangladesh in the west as well as in the south. The district forms part of the Himalayan Terai of West Bengal.

A geopolitical curiosity is that there are 92 Bangladeshi exclaves, with a total area of 47.7 km² in Cooch-Behar. Similarly, there are 106 Indian exclaves inside Bangladesh, with a total area of 69.5 km². These were part of the high stake card or chess games centuries ago between two regional kings, the Raja of Cooch Behar and the Maharaja of Rangpur.[4]

Twenty-one of the Bangladeshi exclaves are within Indian exclaves, and three of the Indian exclaves are within Bangladeshi exclaves. The largest Indian exclave, Balapara Khagrabari, surrounds a Bangladeshi exclave, Upanchowki Bhajni, which itself surrounds an Indian exclave called Dahala Khagrabari, of less than one hectare (link to external map here [5]). See also Indo-Bangladesh enclaves.

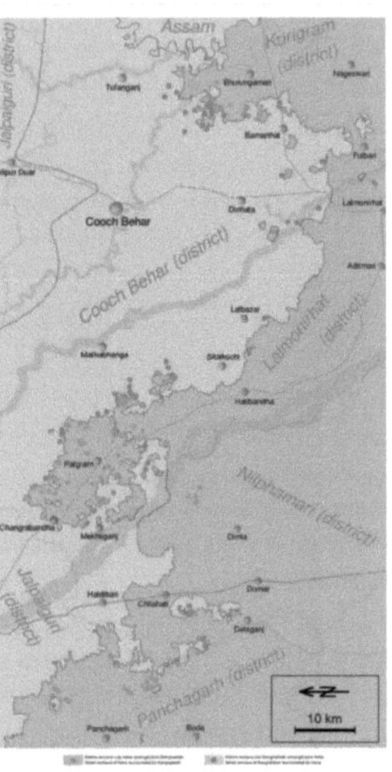

Exclaves of India and Bangladesh in and around Cooch Behar district

Soil

Being the district near the Eastern Himalayan foothills, after rains in the catchment area of each of the rivers generally attain strong current and flood the adjacent area. The turbulent water carries sand, silt, pebbles which causes many problems in productivity as well as hydrology. The soil is formed by alluvial deposits and is acidic in nature. It is friable loam to sandy loam ranging in depth from 0.15 to 1 meter. The soil has a low level of nitrogen while potassium and phosphorus levels are medium. Deficiency of zinc, calcium, magnesium and sulphur is quite high.

Rivers and topography

Cooch Behar is a flat country with a slight southeastern slope along which the main rivers of the district flow. Most of the highland areas are in the Sitalkuchi region and most of the low-lying lands lie in Dinhata region.

The rivers in the district of Cooch Behar generally flow from northwest to southeast. Six rivers that cut through the district are the Teesta, Jaldhaka, Torsha, Kaljani, Raidak, Gadadhar and Ghargharia.

Climate

The district of Cooch Behar has a moderate type of climate characterised by heavy rainfall during the monsoon and slight rainfall in the month of October to mid-November. The district does not have high temperatures at any time of the year. The summer season is from April to May with April being the hottest month with mean daily maximum of 32.5 °C and mean daily minimum of 20.2 °C. The winter season lasts from late November to February, with January being the coldest month with temperature ranging from 10.4 °C to 24.1 °C. The recorded temperature minimum is 3.9 °C and respective recorded maximum is 39.9 °C. The atmosphere is highly humid throughout the year, except the period from February to May, when the relative humidity is as low as 50 to 70%. The rainy season lasts from June to September. The district's average annual rainfall is 3 201 mm.

Economy

Agriculture

The agricultural area of Cooch Behar is 2530.63 square kilometers. The dominant agricultural products of Cooch Behar district are jute and tobacco. Paddy rice is also grown before and after the rainy season. Common plantation crops are arecanut, coconut and black pepper. Vegetable, mustard plant, and potato cultivation are increasing. In order to support agriculture, special programs have been taken for the production of sunflowers, maize and groundnuts. Revolutionary methods are being used in Boro paddy and potato cultivation. But due to nonadoption of modern technology, a large number of farmers still depend on traditional technology. Only 33% of the potentially cultivable land is developed for irrigation. In Kharif, the area of production of vegetables and other crops is much less. The ovine breed in the region originates from Tibet and was brought to the plains of West Bengal by traders. The trade between Tibetan traders and traders from the plains of Bengal took place from the region. The sheep along with other items of trade were transported to a place known as Bhot Patti (situated in Maynaguri Block of Jalpaiguri District). The major trading occurred at a place known as Rangpur, situated now in Bangladesh. The goods were exchanged and the sheep were also taken to plains of Bengal by the returning traders, the animals were given to the farmers of Sunderban region for rearing and bringing them back to their health. The sheep were used for their meat by the Europeans during the colonial era. They preferred mutton over Chevon so sheep meat was in great demand. A single consignment of the sheep were transported to Australia in the late 18th century when the Australian colony was being settled. The consignment was shipped from the port of Fulta near Kolkata. However, the sheep were not preferred by the settlers as their size was small and wool quality too was inferior. The breed Booroola Merino of Australia are the descendents of the same sheep.

Divisions

Administrative divisions

The district comprises five subdivisions: Cooch Behar Sadar, Dinhata, Mathabhanga, Mekhliganj and Tufanganj. Cooch Behar Sadar consists of Cooch Behar municipality and two community development blocs: Cooch Behar–I and Cooch Behar–II. Dinhata subdivision consists of Dinhata municipality and three community development blocs: Dinhata–I, Dinhata–II and Sitai. Mathabhanga subdivision consists of Mathabhanga municipality and three community development blocs: Sitalkuchi, Mathabhanga–I and Mathabhanga–II. Mekhliganj subdivision consists of Mekhliganj municipality and Haldibari municipality and two community development blocs: Mekhliganj and Haldibari. Tufanganj subdivision consists of Tufanganj municipality and two community development blocs: Tufanganj–I and Tufanganj–II.[6] Cooch Behar is the district headquarters. There are 11 police stations, 12 development blocks, 6 municipalities and 128 gram panchayats in this district.[7]

Other than municipality area, each subdivision contains community development blocs which in turn are divided into rural areas and census towns.[8] In total there are 10 urban units: 6 municipalities and 4 census towns. Cooch Behar, Kharimala Khagrabari and Guriahati are part of a urban agglomeration. Dinhata and Bhangri Pratham Khanda also form an urban agglomeration together.

Cooch Behar Sadar subdivision

- Cooch Behar: municipality
- Cooch Behar I (Community development block) consists of rural areas with 15 gram panchayats and two census towns: Kharimala Khagrabari and Guriahati. Block headquarter is in Dhaluabari.
- Cooch Behar II (Community development block) consists of rural areas with 13 gram panchayats and one census town: Khagrabari. Block headquarter is in Pundibari.

Dinhata subdivision

- Dinhata: municipality
- Dinhata I (Community development block) consists of rural areas with 16 gram panchayats and one census town: Bhangri Pratham Khanda. Block headquarter is in Dinhata.
- Dinhata II (Community development block) consists of rural areas only with 12 gram panchayats. Block headquarter is in Sahebganj.
- Sitai (Community development block) consists of rural areas only with 5 gram panchayats. Block headquarter is in Sitai.

Mathabhanga subdivision

- Mathabhanga: municipality
- Sitalkuchi (Community development block) consists of rural areas only with 8 gram panchayats. Block headquarter is in Sitalkuchi.
- Mathabhanga I (Community development block) consists of rural areas only with 10 gram panchayats. Block headquarter is in Sikarpur.
- Mathabhanga II (Community development block) consists of rural areas only with 19 gram panchayats. Block headquarter is in Mathabhanga.

Mekhliganj subdivision

- Mekhliganj: municipality
- Haldibari: municipality
- Mekhliganj (Community development block) consists of rural areas only with 8 gram panchayats. Block headquarter is in Changrabandha.
- Haldibari (Community development block) consists of rural areas only with 6 gram panchayats. Block headquarter is in Haldibari.

Tufanganj subdivision

- Tufanganj: municipality
- Tufanganj I (Community development block) consists of rural areas only with 14 gram panchayats. Block headquarter is in Tufanganj.
- Tufanganj II (Community development block) consists of rural areas only with 11 gram panchayats. Block headquarter is in Baxirhat.

Assembly constituencies

The district is divided into 9 assembly constituencies:[9]

1. Mekliganj (SC) (assembly constituency no. 1),
2. Sitalkuchi (SC) (assembly constituency no. 2),
3. Mathabhanga (SC) (assembly constituency no. 3),
4. Cooch Behar North (assembly constituency no. 4),
5. Cooch Behar West (assembly constituency no. 5),
6. Sitai (assembly constituency no. 6),
7. Dinhata (assembly constituency no. 7),
8. Natabari (assembly constituency no. 8) and
9. Tufanganj (SC) (assembly constituency no. 9).

Mekliganj, Sitalkuchi, Mathabhanga and Tufanganj constituencies are reserved for Scheduled Castes (SC) candidates. Mekhliganj constituency is part of Jalpaiguri (Lok Sabha constituency), which also contains six assembly segments from Jalpaiguri district. Sitalkuchi, Mathabhanga, Cooch Behar North, Cooch Behar West, Sitai, Dinhata and Natabari constituencies form the Cooch Behar (Lok Sabha constituency), which is reserved for Scheduled Castes (SC). Tufanganj constituency is part of Alipurduars (Lok Sabha constituency), which also contains six assembly segments from Jalpaiguri district.

Impact of delimitation of constituencies

As per order of the Delimitation Commission in respect of the delimitation of constituencies in the West Bengal, the district will be divided into 9 assembly constituencies:[10]

1. Mekliganj (SC) (assembly constituency no. 1),
2. Mathabhanga (SC) (assembly constituency no. 2),
3. Cooch Behar Uttar (SC) (assembly constituency no. 3),
4. Cooch Behar Dakshin (assembly constituency no. 4),
5. Sitalkuchi (SC) (assembly constituency no. 5),
6. Sitai (SC) (assembly constituency no. 6),
7. Dinhata (assembly constituency no. 7),
8. Natabari (assembly constituency no. 8) and
9. Tufanganj (assembly constituency no. 9).

Mekliganj, Mathabhanga, Cooch Behar Uttar, Sitalkuchi and Sitai constituencies will be reserved for Scheduled Castes (SC) candidates. Mekhliganj constituency will remain part of Jalpaiguri (Lok Sabha constituency), which will also contain six assembly segments from Jalpaiguri district. Mathabhanga, Cooch Behar Uttar, Cooch Behar Dakshin, Sitalkuchi, Sitai, Dinhata and Natabari constituencies will continue to form the Cooch Behar (Lok Sabha constituency), which will be reserved for Scheduled Castes (SC). Tufanganj constituency will remain a part of Alipurduars (Lok Sabha constituency), which will also contain six assembly segments from Jalpaiguri district.

Culture

Tourism

- **Cooch Behar Palace** (Rajbari): Built in the classical European style of Italian Renaissance on the lines of Buckingham Palace in 1887. A recently constructed museum in the rooms of the Palace has added glory to the Royal structure. The vast lawn and beautiful landscaping of the garden have made it more beautiful. It is a must visit.

Cooch Behar Palace

- **Madan Mohan Temple**: Situated in the heart of the Cooch Behar town. Constructed by Maharaja Nripendra Narayan during 1885 to 1889. A divine structure, deities include Madan Mohan the *kul-devata* of the Koch Dynasty, Ma Tara and Ma Bhavani. The annual Rash Mela is held here in November.

- **Rajpat Mound**: A protected monument by the Archaeological Survey of India (ASI). Situated about 35 km from Cooch Behar Town. One can see the remains of a palace and some excavated artifacts and statues.

- **Baneshwar Shiv Temple**: Situated at a distance of about 10 km to the North of Cooch Behar town, the temple has a 'Shivalinga' 10 feet below the plinth level. There is a big pond within the temple campus having a large number of tortoise. Some of the tortoises are very old and big in size. At Siva Chaturdashi a big mela is held here for a week.

- **Madhupur Dham**: Situated about 10 km west from Cooch Behar Town. In 1489, Shankaradeva performed his last journey to Cooch Behar when Maharaja Nar Narayan requested him to preach the teachings of the neo-Vaishnava cult. It was in his honour that the Madhupur Dham was built in the 16th century. This place has a special significance for the devotees of Acharya Shankaradeva.

- **Kamteswari Temple**: Situated at a distance of about 35 km west of Cooch Behar Town, the original temple is now destroyed. The present temple has been established by Maharaja Pran Narayan in 1665 The throne of Devi is situated here. Beside the main temple 2 smaller temples also exist/ at the gate a 'Tarakeswar Sivalinga' exists.

Rasikbeel

- **Sagardighi**: Situated in the Cooch Behar Town itself. The huge tank was excavated by Maharaja Hitendra Narayan. It is a popular rendezvous in the evening, surrounded by heritage buildings including Victor House and a War Memorial where a tank is kept. During winter months one can spot migratory birds on the water surface and the nearby trees.

Sagar Dighi

- **Rasikbil**: It is situated about 42 km from Cooch Behar Town. A recognized bird sanctuary. It has a deer park and a recently built aquarium where fishes, turtles, seven nos. of leopards, Peafowl are kept. You may spot Chinese Fishing Nets on the way to Rasikbil. Rasikbeel is a complex of wet land, the name of important water bodies are Bochamari beel, Rasik beel, Batikata Beel & raichangmari beel. In Bengali beel means large water body. The main migratory bird spp found in this wet land are Lesser Whistling Teal, Common Teal, Cotton Teal, Dapchick, Bronze winged Jacana, Pheasant Tailed Janacana, Shoveler, Barheaded goose, White Eyed Poacherd etc. Except this a lot of other aquatic bird like small & large Cormorant, four spp. of Kingfisher, open bill stork etc. are found. The area of water doby complex is 178 hec. The whole area comes under protected forest & managed by Coochbehar Forest Division. In recent past (Jan, 09) a beautiful watch tower of 70 feet height was constructed by Coochbehar Forest Division. There is a min zoo at Rasikbill, the zoo is recognised by Central Zoo Authority, Govt. of India. There are Tortoise, Gharial, Leopard, Spotted deer, Peafowl and other birds in the zoo. In 2009, Coochbehar Division in collaboration with Zoological Survey of India conducted bird census in the wetland complex, 66 species of birds were recorded.

Rasikbeel Watch Tower

- **Rasomati Ecotourism complex**- This ecotourism complex is recently developed by Coochbehar Forest Division. The main attraction is the Rasomati Jheel (Water body) which herbours lots of residential & migratory birds. There is a picnic spot with paddle boating facility for tourist. A six km. long Jungle safari is also major tourist attraction. For observation a tower of height 56 feet has been constructed. The spot is located in the Patlakhawa forest which was a game reserve of the king of Coochbehar.

- **Kholta Ecotourism Spot**- The spot is on Coochbehar Aliporeduar Road, 20 km away from Coochbehar, Recently (Feb-09) Developed by Coochbehar Forest Division, The children park, Deer park (Sambar & Spotted Deer)& Toy train is the major tourist attraction. The spot is surrounded by Araikumari riverlet, there is old teak plantation created by the king of Coochbehar.

Apart from these other tourist spots are:

- Eco Heritage Park
- Nipendra Narayan Park
- Brahmo Mandir

- Ranir Bagan
- Baradebi Bari
- Siddheswari Kali Bari
- Dangar Ayee Temple
- Siddhanath Siva temple, Dhaluabari
- Madan Mohan temple, Mathabhanga

Demographics

According to the 2011 census Cooch Behar district has a population of 2,822,780,[3] roughly equal to the nation of Jamaica[11] or the US state of Kansas.[12] This gives it a ranking of 136th in India (out of a total of 640).[3] The district has a population density of 833 inhabitants per square kilometre (**unknown operator: u'strong'** /sq mi) .[3] Its population growth rate over the decade 2001-2011 was 13.86 %.[3] Koch Bihar has a sex ratio of 942 females for every 1000 males,[3] and a literacy rate of 75.49 %.[3]

Flora and fauna

The flora here includes among others palms, bamboos, creepers, ferns, orchids, aquatic plants, fungi, timber, grass, vegetable and fruit trees.

In absence of large forest area in the district, except at Patlakhawa, not many species of animal are found though there are many wildlife sanctuaries, national parks and animal reserves in the neighboring Jalpaiguri district and Alipurduar subdivision of Jalpaiguri which are not very far from the district.

In 1976 Cooch Behar district became home to the Jaldapara Wildlife Sanctuary, which has an area of 217 km^2 (**unknown operator: u'strong'** sq mi).[13] It shares the park with Jalpaiguri district.[13]

Education

Educational Facilities

Primary Schools - 1805

High Schools - 120

Higher Secondary Schools - 61

High Madrasa - 5

Senior Madrasa - 2

Junior High School - 60

Junior High Madrasa - 16

Kendriya vidyalaya - 1

Engineering / Technical Schools - 2

Professional & Technical Schools - 16

General College - 9

Blind School - 1

Libraries - 110

Cooch Behar district has an Agricultural University named Uttar Banga Krishi Viswavidyalaya at Pundibari about 15 km from Cooch Behar Town. Apart from those Government schools there are a few privately aided schools mostly ICSE, ISC and CBSE boards.

See also

- List of enclaves and exclaves
- Indo-Bangladesh enclaves

References

[1] "District-specific Literates and Literacy Rates, 2001" (http://www.educationforallinindia.com/page157.html). Registrar General, India, Ministry of Home Affairs. . Retrieved 2010-10-10.

[2] http://www.coochbehar.gov.in

[3] "District Census 2011" (http://www.census2011.co.in/district.php). Census2011.co.in. 2011. . Retrieved 2011-09-30.

[4] "A Great Divide" (http://www.time.com/time/magazine/article/0,9171,1877200-4,00.html). *Time*. 2009-02-05. .

[5] http://geosite.jankrogh.com/enklaver/CoochBehar_Annotated.jpg

[6] "Directory of District, Sub division, Panchayat Samiti/ Block and Gram Panchayats in West Bengal, March 2008" (http://wbdemo5.nic.in/ writereaddata/Directoryof_District_Block_GPs(RevisedMarch-2008).doc). *West Bengal*. National Informatics Centre, India. 2008-03-19. . Retrieved 2008-11-08.

[7] "District Administration of Cooch Behar" (http://coochbehar.nic.in/Htmfiles/District_Administration.html). Official website of the Cooch Behar district. . Retrieved 2008-11-08.

[8] "Population, Decadal Growth Rate, Density and General Sex Ratio by Residence and Sex, West Bengal/ District/ Sub District, 1991 and 2001" (http://www.wbcensus.gov.in/DataTables/02/Table4_3.htm). *West Bengal*. Directorate of census operations. . Retrieved 2008-11-08.

[9] "General election to the Legislative Assembly, 2001 – List of Parliamentary and Assembly Constituencies" (http://archive.eci.gov.in/ se2001/background/S25/WB_ACPC.pdf) (PDF). *West Bengal*. Election Commission of India. . Retrieved 2008-11-16.

[10] "Press Note, Delimitation Commission" (http://www.wbgov.com/e-gov/English/DELIMITATION.pdf) (PDF). *Assembly Constituencies in West Bengal*. Delimitation Commission. . Retrieved 2008-11-16.

[11] US Directorate of Intelligence. "Country Comparison:Population" (https://www.cia.gov/library/publications/the-world-factbook/ rankorder/2119rank.html). . Retrieved 2011-10-01. "Jamaica 2,868,380 July 2011 est"

[12] "2010 Resident Population Data" (http://2010.census.gov/2010census/data/apportionment-pop-text.php). U. S. Census Bureau. . Retrieved 2011-09-30. "Kansas 2,853,118"

[13] Indian Ministry of Forests and Environment. "Protected areas: Sikkim" (http://oldwww.wii.gov.in/envis/envis_pa_network/index.htm). . Retrieved September 25, 2011.

- Moore, Lucy (2004) *Maharanis: The Extraordinary Tale Of Four Indian Queens And Their Journey From Purdah To Parliament*, Penguin, ISBN 0-670-03368-5

External links

- Website of the district administration of Cooch Behar (http://www.coochbehar.gov.in)
- Crime Report (http://ncrb.nic.in/crime2005/cii-2005/Table 1.14.pdf)
- Link to external map (http://geosite.jankrogh.com/enklaver/CoochBehar_Annotated.jpg)

West_Bengal

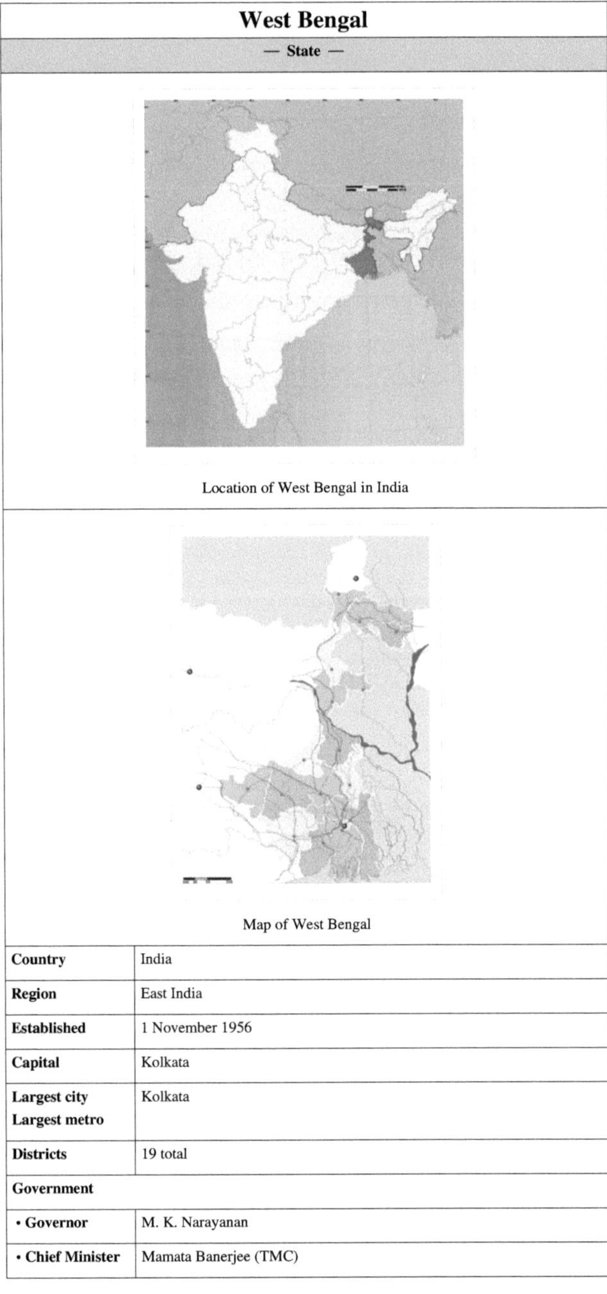

<table>
<tr><td colspan="2" align="center">West Bengal</td></tr>
<tr><td colspan="2" align="center">— State —</td></tr>
<tr><td colspan="2" align="center">Location of West Bengal in India</td></tr>
<tr><td colspan="2" align="center">Map of West Bengal</td></tr>
<tr><td>Country</td><td>India</td></tr>
<tr><td>Region</td><td>East India</td></tr>
<tr><td>Established</td><td>1 November 1956</td></tr>
<tr><td>Capital</td><td>Kolkata</td></tr>
<tr><td>Largest city
Largest metro</td><td>Kolkata</td></tr>
<tr><td>Districts</td><td>19 total</td></tr>
<tr><td>Government</td><td></td></tr>
<tr><td>• Governor</td><td>M. K. Narayanan</td></tr>
<tr><td>• Chief Minister</td><td>Mamata Banerjee (TMC)</td></tr>
</table>

• Legislature	Unicameral (295 • seats)
Area	
• Total	88752 km^2 (**unknown operator: u'strong'** sq mi)
Area rank	13th
Population (2011)[1]	
• Total	91347736
• Rank	4th
• Density	**unknown operator: u'strong'/km^2 (unknown operator: u'strong'/sq mi)**
Time zone	IST (UTC+05:30)
ISO 3166 code	IN-WB
HDI	▼ 0.625 (medium)
HDI rank	19th (2005)
Literacy	77.08%[2]
Official languages	Bengali · English
Website	westbengal.gov.in [3]
294 elected, 1 nominated	

West Bengal /bɛŋˈgɔːl/ (proposed new English name: *Paschim Banga*[4]) is a state in the eastern region of India and is the nation's fourth-most populous.[5] It is also the seventh-most populous sub-national entity in the world, with over 91 million inhabitants.[5] Covering a total area of 34267 sq mi (**unknown operator: u'strong'** km^2), it is bordered by the countries of Nepal, Bhutan, and Bangladesh, and the Indian states of Orissa, Jharkhand, Bihar, Sikkim, and Assam. The state capital is Kolkata (formerly *Calcutta*). West Bengal encompasses two broad natural regions: the Gangetic Plain in the south and the sub-Himalayan and Himalayan area in the north.

In the 3rd century BC, the broader region of Bengal was conquered by the emperor Ashoka. In the 4th century AD, it was absorbed into the Gupta Empire. From the 13th century onward, the region was ruled by several sultans, powerful Hindu states and Baro-Bhuyan landlords, until the beginning of British rule in the 18th century. The British East India Company cemented their hold on the region following the Battle of Plassey in 1757, and the city of Calcutta (now known as Kolkata) served for many years as the capital of British India. The early and prolonged exposure to British administration resulted in expansion of Western education, culminating in development in science, institutional education, and social reforms of the region, including what became known as the Bengal Renaissance. A hotbed of the Indian independence movement through the early 20th century, Bengal was divided in 1947 along religious lines into two separate entities: West Bengal − a state of India − and East Bengal, which initially joined the new nation of Pakistan, before becoming part of modern-day Bangladesh in 1971.

A major agricultural producer, West Bengal is the sixth-largest contributor to India's net domestic product.[6] Noted for its political activism, the state was ruled by democratically elected communist government for three decades. West Bengal is noted for its cultural activities, with the state capital Kolkata earning the sobriquet "cultural capital of India". The state's cultural heritage, besides folk culture, ranges from stalwarts in literature including Nobel-laureate Rabindranath Tagore to scores of musicians, film-makers and artist. West Bengal is also distinct from most other Indian states in its appreciation and practice of playing soccer besides the national favourite sport cricket.[7] [8] [9]

Etymology

The name of *Bengal*, or *Bangla*, is of unknown origins. Many theories have been formulated to explain the origin of the word "Bengal" or "Bangla". One theory suggests that the word derives from Dravidian tribes of 1000 B.C present at that time.[10] The word might have been derived from the ancient kingdom of *Vanga*, or *Banga*. Although some early Sanskrit literature mentions the name, the region's early history is obscure. The region was part of Mauryan empire under Ashoka in the 3rd century BCE.

History

Stone age tools dating back 20,000 years have been excavated in the state.[11] Remnants of civilization in the greater Bengal region date back four thousand years,[12] when the region was settled by Dravidian, Tibeto-Burman, and Austro-Asiatic peoples. The region was a part of the Vanga Kingdom, one of ancient kingdoms of Epic India. The kingdom of Magadha was formed in 7th century BC, consisting of the Bihar and Bengal regions. It was one of the four main kingdoms of India at the time of Mahavira and the Buddha, and consisted of several *Janapadas*.[13] During the rule of Maurya dynasty, the Magadha Empire extended over nearly all of South Asia, including Afghanistan and parts of Persia under Ashoka the Great in the 3rd century BC.

One of the earliest foreign references to Bengal is a mention of a land named Gangaridai by the Ancient Greeks around 100 BC. The word is speculated to have come from *Gangahrd* (Land with the Ganges in its heart) in reference to an area in Bengal.[14] Bengal had overseas trade relations with Suvarnabhumi (Burma, Lower Thailand, Lower Malay Peninsula, and the Sumatra).[15] According to Mahavamsa, Vijaya Singha, a Vanga prince, conquered Lanka (modern day Sri Lanka) and gave the name "Sinhala" to the country.[16]

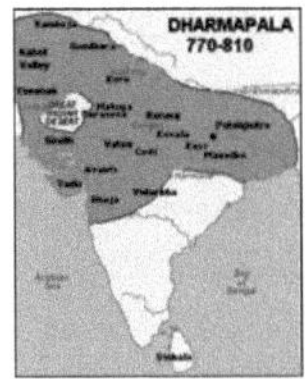

Pala Empire under Dharmapala   Pala Empire under Devapala

From the 3rd to the 6th centuries AD, the kingdom of Magadha served as the seat of the Gupta Empire. The first recorded independent king of Bengal was Shashanka, reigning around early 7th century.[17] After a period of anarchy, the Buddhist Pala dynasty ruled the region for four hundred years, followed by a shorter reign of the Hindu Sena dynasty. Islam made its first appearance in Bengal during the 12th century when Sufi missionaries arrived. Later, occasional Muslim raiders reinforced the process of conversion by building mosques, madrassas and Sufi Khanqah. Between 1202 and 1206, Bakhtiar Khilji, a military commander from the Delhi Sultanate, overran Bihar and Bengal as far east as Rangpur, Bogra and the Brahmaputra River. Although he failed to bring Bengal under his control, the expedition managed to defeat Lakshman Sen and his two sons moved to a place then called Vikramapur (present-day Munshiganj District), where their diminished dominion lasted until the late 13th century.

During the 14th century, the former kingdom became known as the Sultanate of Bengal, ruled intermittently with the Sultanate of Delhi as well as powerful Hindu states and land-lords-Baro-Bhuyans. The Hindu Deva Kingdom ruled over eastern Bengal after the collapse Sena Empire. The Sultanate of Bengal was interrupted by an uprising by the Hindus under Raja Ganesha. The Ganesha dynasty began in 1414, but his successors converted to Islam. Bengal came once more under the control of Delhi as the Mughals conquered it in 1576. There were several independent Hindu states established in Bengal during the Mughal period like those of Maharaja Pratap Aditya of Jessore and Raja Sitaram Ray of Burdwan. These kingdoms contributed greatly to the economic and cultural landscape of Bengal. Extensive land reclamations in forested and marshy areas were carried out and trade as well as commerce were highly encouraged. These kingdoms also helped introduce new music, painting, dancing and sculpture into Bengali art-forms as well as many temples were constructed during this period. Militarily, they served as bulwarks against Portuguese and Burmese attacks. Koch Bihar Kingdom in the northern Bengal, flourished during the period of 16th and the 17th centuries as well as weathered the Mughals and survived till the advent of the British.

Raja Ram Mohan Roy is widely regarded as the "Father of the Bengal Renaissance".

European traders arrived late in the fifteenth century. Their influence grew until the British East India Company gained taxation rights in Bengal *subah*, or province, following the Battle of Plassey in 1757, when Siraj ud-Daulah, the last independent Nawab, was defeated by the British.[18] The Bengal Presidency was established by 1765, eventually including all British territories north of the Central Provinces (now Madhya Pradesh), from the mouths of the Ganges and the Brahmaputra to the Himalayas and the Punjab. The Bengal famine of 1770 claimed millions of lives.[19] Calcutta was named the capital of British India in 1772. The Bengal Renaissance and Brahmo Samaj socio-cultural reform movements had great impact on the cultural and economic life of Bengal. The failed Indian rebellion of 1857 started near Calcutta and resulted in transfer of authority to the British Crown, administered by the Viceroy of India.[20] Between 1905 and 1911, an abortive attempt was made to divide the province of Bengal into two zones.[21] Bengal suffered from the Great Bengal famine in 1943 that claimed 3 million lives.[22]

Bengal played a major role in the Indian independence movement, in which revolutionary groups such as *Anushilan Samiti* and *Jugantar* were dominant. Armed attempts against the British Raj from Bengal reached a climax when Subhash Chandra Bose led the Indian National Army from Southeast Asia against the British. When India gained independence in 1947, Bengal was partitioned along religious lines. The western part went to India (and was named West Bengal) while the eastern part joined Pakistan as a province called East Bengal (later renamed East Pakistan, giving rise to independent Bangladesh in 1971).[23] Both West and East Bengal suffered from large refugee influx during the partition in 1947, leading to the political unrests later on. The partition of Bengal entailed the greatest exodus of people in Human History. Millions of Hindus migrated from East Pakistan to India and thousands of Muslims too went across the borders to East Pakistan. Because of the immigration of the refugees, there occurred the crisis of land and food in West Bengal; and such condition remained in long duration for more than three decades.The politics of West Bengal since the partition in 1947 developed round the nucleus of refugee

Subhash Chandra Bose was one of the most prominent Bengali freedom fighters in India's struggle for independence against the British Raj.

problem. Both the Rightists and the Leftists in the Politics of West Bengal have not yet become free from the socio-economic conditions created by the partition of Bengal. These conditions as have remained unresolved in some twisted forms have given birth to local socio-economic, political and ethnic movements.[24]

In 1950, the Princely State of Cooch Behar merged with West Bengal after King Jagaddipendra Narayan signed the Instrument of Accession with India.[25] In 1955, the former French enclave of Chandannagar, which had passed into Indian control after 1950, was integrated into West Bengal; portions of Bihar were subsequently merged with West Bengal.

During the 1970s and 1980s, severe power shortages, strikes and a violent Marxist-Naxalite movement damaged much of the state's infrastructure, leading to a period of economic stagnation. The Bangladesh Liberation War of 1971 resulted in the influx of millions of refugees to West Bengal, causing significant strains on its infrastructure.[26] The 1974 smallpox epidemic killed thousands. West Bengal politics underwent a major change when the Left Front won the 1977 assembly election, defeating the incumbent Indian National Congress. The Left Front, led by Communist Party of India (Marxist), governed for the state for the subsequent three decades.[27]

The state's economic recovery gathered momentum after economic reforms were introduced in the mid-1990s by the central government, aided by the advent of information technology and IT-enabled services. As of 2007, armed activists have been conducting minor terrorist attacks in some parts of the state,[28] [29] while clashes with the administration are taking place at several sensitive places over the issue of industrial land acquisition.[30] [31] Although the state's GDP has risen significantly since the 1990s, West Bengal has remained affected by political instability and bad governance.[32] The state continues to suffer from regular bandhs (strikes),[33] [34] a low Human Development Index level,[35] substandard healthcare services,[36] [37] a lack of socio-economic development,[38] poor infrastructure,[39] [40] political corruption and civil violence.[41] [42]

Geography and climate

West Bengal is on the eastern bottleneck of India, stretching from the Himalayas in the north to the Bay of Bengal in the south. The state has a total area of 88752 square kilometres (**unknown operator: u'strong'** sq mi).[1] The Darjeeling Himalayan hill region in the northern extreme of the state belongs to the eastern Himalaya. This region contains Sandakfu (3636 m or **unknown operator: u'strong'** ft)—the highest peak of the state.[43] The narrow Terai region separates this region from the plains, which in turn transitions into the Ganges delta towards the south. The Rarh region intervenes between the Ganges delta in the east and the western plateau and high lands. A small coastal region is on the extreme south, while the Sundarbans mangrove forests form a remarkable geographical landmark at the Ganges delta.

Many areas remain flooded during the heavy rains brought by monsoon

The Ganges is the main river, which divides in West Bengal. One branch enters Bangladesh as the *Padma* or *Pôdda*, while the other flows through West Bengal as the Bhagirathi River and Hooghly River. The Farakka barrage over Ganges feeds the Hooghly branch of the river by a feeder canal, and its water flow management has been a source of lingering dispute between India and Bangladesh.[44] The Teesta, Torsa, Jaldhaka and Mahananda rivers are in the northern hilly region. The western plateau region has rivers such as the Damodar,

Ajay and Kangsabati. The Ganges delta and the Sundarbans area have numerous rivers and creeks. Pollution of the Ganges from indiscriminate waste dumped into the river is a major problem.[45] Damodar, another tributary of the Ganges and once known as the "Sorrow of Bengal" (due to its frequent floods), has several dams under the Damodar Valley Project. At least nine districts in the state suffer from arsenic contamination of groundwater, and an estimated 8.7 million people drink water containing arsenic above the World Health Organisation recommended limit of 10 µg/L.[46]

National Highway 31A winds along the banks of the Teesta River near Kalimpong, in the Darjeeling Himalayan hill region.

West Bengal's climate varies from tropical savanna in the southern portions to humid subtropical in the north. The main seasons are summer, rainy season, a short autumn, and winter. While the summer in the delta region is noted for excessive humidity, the western highlands experience a dry summer like northern India, with the highest day temperature ranging from 38 °C (**unknown operator: u'strong'** °F) to 45 °C (**unknown operator: u'strong'** °F).[47] At nights, a cool southerly breeze carries moisture from the Bay of Bengal. In early summer brief squalls and thunderstorms known as *Kalbaisakhi*, or Nor'westers, often occur.[48] West Bengal receives the Bay of Bengal branch of the Indian ocean monsoon that moves in a northwest direction. Monsoons bring rain to the whole state from June to September. Heavy rainfall of above 250 cm is observed in the Darjeeling, Jalpaiguri and Cooch Behar district. During the arrival of the monsoons, low pressure in the Bay of Bengal region often leads to the occurrence of storms in the coastal areas. Winter (December–January) is mild over the plains with average minimum temperatures of 15 °C (**unknown operator: u'strong'** °F).[47] A cold and dry northern wind blows in the winter, substantially lowering the humidity level. However, the Darjeeling Himalayan Hill region experiences a harsh winter, with occasional snowfall at places.

Flora and fauna

State symbols of West Bengal

Union day	August 18 (Day of accession to India)	
State animal	Fishing cat[49]	
State bird	White-throated Kingfisher	
State tree	Devil Tree[49]	
State flower	Night-flowering Jasmine[49]	

A Bengal tiger.

Sal trees in the Arabari forest in West Midnapur.

As of 2009, recorded forest area in the state is 11879 km^2 (**unknown operator: u'strong'** sq mi) which is 13.38% of the state's geographical area,[50] compared to the national average of 21.02%.[51] [52] Reserves, protected and unclassed forests constitute 59.4%, 31.8% and 8.9%, respectively, of the forest area.[50] Part of the world's largest mangrove forest, the Sundarbans, is located in southern West Bengal.[53]

From a phytogeographic viewpoint, the southern part of West Bengal can be divided into two regions: the Gangetic plain and the littoral mangrove forests of the Sundarbans.[54] The alluvial soil of the Gangetic plain, compounded with favorable rainfall, make this region especially fertile.[54] Much of the vegetation of the western part of the state shares floristic similarities with the plants of the Chota Nagpur plateau in the adjoining state of Jharkhand.[54] The predominant commercial tree species is *Shorea robusta*, commonly known as the Sal tree. The coastal region of Purba Medinipur exhibits coastal vegetation; the predominant tree is the *Casuarina*. A notable tree from the Sundarbans is the ubiquitous *sundari* (*Heritiera fomes*), from which the forest gets its name.[55]

The distribution of vegetation in northern West Bengal is dictated by elevation and precipitation. For example, the foothills of the Himalayas, the *Dooars*, are densely wooded with Sal and other tropical evergreen trees.[56] However, above an elevation of 1000 metres (**unknown operator: u'strong'** ft), the forest becomes predominantly subtropical. In Darjeeling, which is above 1500 metres (**unknown operator: u'strong'** ft), temperate-forest trees such as oaks, conifers, and rhododendrons predominate.[56]

West Bengal has 3.26% of its geographical area under protected areas comprising 15 wildlife sanctuaries and 5 national parks[50] — Sundarbans National Park, Buxa Tiger Reserve, Gorumara National Park, Neora Valley National Park and Singalila National Park. Extant wildlife include Indian rhinoceros, Indian elephant, deer, bison, leopard, gaur, tiger, and crocodiles, as well as many bird species. Migratory birds come to the state during the winter.[57] The high-altitude forests of Singalila National Park shelter barking deer, red panda, chinkara, takin, serow, pangolin, minivet and Kalij pheasants. The Sundarbans are noted for a reserve project conserving the endangered Bengal tiger, although the forest hosts many other endangered species, such as the Gangetic dolphin, river terrapin and estuarine crocodile.[58] The mangrove forest also acts as a natural fish nursery, supporting coastal fishes along the Bay of Bengal.[58] Recognizing its special conservation value, Sundarban area has been declared as a Biosphere Reserve.[50]

Government and politics

West Bengal is governed through a parliamentary system of representative democracy, a feature the state shares with other Indian states. Universal suffrage is granted to residents. There are two branches of government. The legislature, the West Bengal Legislative Assembly, consists of elected members and special office bearers such as the Speaker and Deputy Speaker, that are elected by the members. Assembly meetings are presided over by the Speaker or the Deputy Speaker in the Speaker's absence. The judiciary is composed of the Calcutta High Court and a system of lower courts. Executive authority is vested in the Council of Ministers headed by the Chief Minister, although the titular head of government is the Governor. The Governor is the head of state appointed by the President of India. The leader of

Calcutta High Court is the highest court in West Bengal

the party or coalition with a majority in the Legislative Assembly is appointed as the Chief Minister by the Governor, and the Council of Ministers are appointed by the Governor on the advice of the Chief Minister. The Council of Ministers reports to the Legislative Assembly. The Assembly is unicameral with 295 Members of the Legislative Assembly, or MLAs,[59] including one nominated from the Anglo-Indian community. Terms of office run for 5 years, unless the Assembly is dissolved prior to the completion of the term. Auxiliary authorities known as *panchayats*, for which local body elections are regularly held, govern local affairs. The state contributes 42 seats to Lok Sabha[60] and 16 seats to Rajya Sabha of the Indian Parliament.[61]

The main players in the regional politics are the All India Trinamool Congress, the Indian National Congress, the Left Front alliance (led by the Communist Party of India (Marxist) or CPI(M)). Following the West Bengal State Assembly Election in 2011, the All India Trinamool Congress and Indian National Congress coalition under Mamata Banerjee of the All India Trinamool Congress was elected to power (getting 225 seats in the legislature).[62] West Bengal was ruled by the Left Front for the 34 years (1977–2011), making it the world's longest-running democratically elected communist government.[27]

Subdivisions

The 19 districts of West Bengal are as listed below.[63]

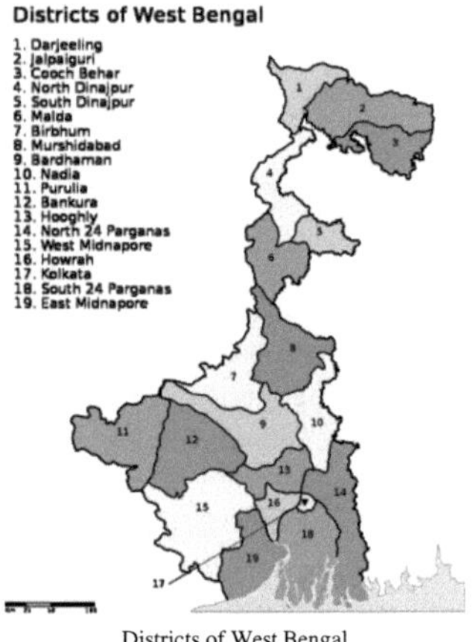

Districts of West Bengal

- Bankura
- Bardhaman
- Birbhum
- Cooch Behar
- Darjeeling
- East Midnapore
- Hooghly
- Howrah
- Jalpaiguri
- Kolkata
- Malda
- Murshidabad
- Nadia
- North 24 Parganas
- Uttar Dinajpur
- Purulia
- South 24 Parganas
- Dakshin Dinajpur
- West Midnapore

Each district is governed by a district collector or district magistrate, appointed either by the Indian Administrative Service or the West Bengal Civil Service.[64] Each district is subdivided into Sub-Divisions, governed by a sub-divisional magistrate, and again into Blocks. Blocks consists of panchayats (village councils) and town municipalities.[63]

The capital and largest city of the state is Kolkata – the third-largest urban agglomeration[65] and the seventh-largest city[66] in India. Asansol is the second largest city & urban agglomeration in West Bengal after Kolkata.[65] Siliguri is an economically important city, strategically located in the northeastern Siliguri Corridor (Chicken's Neck) of India. Other major cities and towns in West Bengal are Howrah, Durgapur, Raniganj, Haldia, Jalpaiguri, Kharagpur, Burdwan, Darjeeling, Midnapore, and Malda.[66]

Economy

Roadside vegetable vendor in a semi-rural area. A large proportion of residents are employed in informal sector.

Net State Domestic Product at Factor Cost at Current Prices (2004–05 Base)[6] figures in crores of Indian Rupees	
Year	Net State Domestic Product
2004–2005	190,073
2005–2006	209,642
2006–2007	238,625
2007–2008	272,166
2008–2009	309,799
2009–2010	366,318

In 2009–10, the tertiary sector of the economy (service industries) was the largest contributor to the gross domestic product of the state, contributing 57.8% of the state domestic product compared to 24% from primary sector (agriculture, forestry, mining) and 18.2% from secondary sector (industrial and manufacturing).[67] :12 Agriculture is the leading occupation in West Bengal. Rice is the state's principal food crop. Rice, potato, jute, sugarcane and wheat are the top five crops of the state.[67] :14 Tea is produced commercially in northern districts; the region is well known for Darjeeling and other high quality teas.[67] :14 State industries are localized in the Kolkata region, the mineral-rich western highlands, and Haldia port region.[68] The Durgapur–Asansol colliery belt is home to a number of major steel plants.[68] Manufacturing industries playing an important economic role are engineering products, electronics, electrical equipment, cables, steel, leather, textiles, jewellery, frigates, automobiles, railway coaches, and wagons. The Durgapur centre has established a number of industries in the areas of tea, sugar, chemicals and fertilizers. Natural resources like tea and jute in and nearby parts has made West Bengal a major centre for the jute and tea industries.

A significant part of the state is economically backward, namely, large parts of six northern districts of Cooch Behar, Darjeeling, Jalpaiguri, Malda, North Dinajpur and South Dinajpur; three western districts of Purulia, Bankura, Birbhum; and the Sundarbans area.[69] Years after independence, West Bengal was still dependent on the central government for meeting its demands for food; food production remained stagnant and the Indian green revolution bypassed the state. However, there has been a significant spurt in food production since the 1980s, and the state now has a surplus of grains.[69] The state's share of total industrial output in India was 9.8% in 1980–81, declining to 5% by 1997–98. However, the service sector has grown at a rate higher than the national rate.[69]

Freshly sown saplings of paddy; in the background are stacks of jute sticks

In terms net state domestic product (NSDP), West Bengal has the sixth largest economy (2009–2010) in India, with an NSDP of 366,318 crore Indian rupees, behind Maharashtra (817,891 crores), Uttar Pradesh (453,020 crores), Andhra Pradesh (426,816 crores), Tamil Nadu (417,716 crores), and Gujarat (370,400 crores).[6] In the period 2004–2005 to 2009–2010, the average gross state domestic product (GSDP) growth rate was 13.9% (calculated in Indian rupee term), lower than 15.5%, the average for all states of the country.[67] :4 The state's per capita GSDP at current prices in 2009–10 was US$ 956.4, improved from US$ 553.7 in 2004–05,[67] :10 but lower than the national per capita GSDP of US$ 1,302.[67] :4 The state's total financial debt stood at ₹ 191835 crore (US$38.27 billion) as of 2011.[70]

The state has promoted foreign direct investment, which has mostly come in the software and electronics fields; Kolkata is becoming a major hub for the Information technology (IT) industry. Rapid industrialisation process has given rise to debate over land acquisition for industry in this agrarian state.[71] NASSCOM–Gartner ranks West Bengal power infrastructure the best in the country.[72] Notably, many corporate companies are now headquartered in Kolkata include ITC Limited, India Government Mint, Kolkata, Haldia Petrochemicals, Exide Industries, Hindustan Motors, Britannia Industries, Bata India, Birla Corporation, CESC Limited, Coal India Limited, Damodar Valley Corporation, PwC India, Peerless Group, United Bank of India, UCO Bank and Allahabad Bank. In 2010s, events such as adoption of "Look East" policy by the government of India, opening of the Nathu La Pass in Sikkim as a border trade-route with China and immense interest in the South East Asian countries to enter the Indian market and invest have put Kolkata in an advantageous position for development in future, particularly with likes of Myanmar, where India needs oil from military regime.[73] [74]

Transport

Kolkata Suburban Railway caters to the commuters of the populous suburbs of Kolkata

As of 2011, the total length of surface road in West Bengal is over 92023 km (**unknown operator: u'strong'** mi);[67] :18 national highways comprise 2578 km (**unknown operator: u'strong'** mi)[75] and state highways 2393 km (**unknown operator: u'strong'** mi).[67] :18 As of 2006, the road density of the state is 103.69 km per 100 km² (166.92 mi per 100 sq mi), higher than the national average of 74.7 km per 100 km² (120 mi per 100 sq mi).[76] Average speed on state highways varies between 40–50 km/h (25–31 mi/h); in villages and towns, speeds are as low as 20–25 km/h (12–16 mi/h) due to the substandard quality of road constructions and low maintenance. As of 2011, the total railway route length is around 4481 km (**unknown operator: u'strong'** mi).[67] :20 Kolkata is the headquarters of two divisions of the Indian Railways—Eastern Railway and South Eastern Railway.[77] The Northeast Frontier Railway (NFR) plies in the northern parts of the state. The Kolkata metro is the country's first underground railway.[78] The Darjeeling Himalayan Railway, part of NFR, is a UNESCO World Heritage Site.[79]

The state's only international airport is Netaji Subhash Chandra Bose International Airport at Dum Dum, Kolkata. Bagdogra airport near Siliguri is another significant airport in the state. Kolkata is a major river-port in eastern India. The Kolkata Port Trust manages both the Kolkata docks and the Haldia docks.[80] There is passenger service to Port Blair on the Andaman and Nicobar Islands and cargo ship service to ports in India and abroad, operated by the Shipping Corporation of India. Ferry is a principal mode of transport in the southern part of the state, especially in the Sundarbans area. Kolkata is the only city in India to have trams as a mode of transport and these are operated by the Calcutta Tramways Company.[81]

Several government-owned organisations operate substandard bus services in the state, including the Calcutta State Transport Corporation, the North Bengal State Transport Corporation, the South Bengal State Transport Corporation, the West Bengal Surface Transport Corporation, and the Calcutta Tramways Company, thus leading to mismanagement. There are also private bus companies. The railway system is a nationalised service without any private investment. Hired forms of transport include metered taxis and auto rickshaws which often ply specific routes in cities. In most of the state, cycle rickshaws, and in Kolkata, hand-pulled rickshaws, are also used for short-distance travel. Large-scale transport accidents in West Bengal are common, particularly the sinking of transport boats and train crashes.[82]

Demographics

Dakshineswar Kali Temple

Tipu Sultan Mosque

According to the provisional results of 2011 national census, West Bengal is the fourth most populous state in India with a population of 91,347,736 (7.55% of India's population).[1] Majority of the population comprises Bengalis.[84] The Marwaris, Bihari and Oriya minority are scattered throughout the state; communities of Sherpas and ethnic Tibetans can be found in Darjeeling Himalayan hill region. Darjeeling district has a large number of Gurkha people of Nepalese origin. West Bengal is home to indigenous tribal *Adivasis* such as Santals, Kol, Koch-Rajbongshi and Toto tribe. There are a small number of ethnic minorities primarily in the state capital, including Chinese, Tamils, Gujaratis, Anglo-Indians, Armenians, Punjabis, and Parsis.[85] India's sole Chinatown is in eastern Kolkata.[86]

Religions in West Bengal[87]	
Religion	Percent
Hindu	72.5%
Muslim	25.2%
Others	2.3%

The official language is Bengali and English.[88] Nepali is the official language in three subdivisions of Darjeeling district.[88] As of 2001, in descending order of number of speakers, the languages of the state are: Bengali, Hindi, Santali, Urdu, Nepali, and Oriya.[88] Languages such as Rajbongshi and Ho are used in some parts of the state.

As of 2001, Hinduism is the principal religion at 72.5% of the total population, while Muslims comprise 25.2% of the total population , being the second-largest community as also the largest minority group; Sikhism, Christianity and other religions make up the remainder.[87] The state contributes 7.8% of India's population.[89] The state's 2001–2011 decennial growth rate was 13.93%,[1] lower than 1991–2001 growth rate of 17.8%,[1] and also lower than the national rate of 17.64%.[90] The gender ratio is 947 females per 1000 males.[90] As of 2011, West Bengal has a population density of 1029 inhabitants per square kilometre (**unknown operator: u'strong'** /sq mi) making it the second-most densely populated state in India, after Bihar.[90]

The literacy rate is 77.08%, higher than the national rate of 74.04%.[91] Data of 1995–1999 showed the life expectancy in the state was 63.4 years, higher than the national value of 61.7 years.[92] About 72% of people live in rural areas. The proportion of people living below the poverty line in 1999–2000 was 31.9%.[69] Scheduled Castes and Tribes form 28.6% and 5.8% of the population respectively in rural areas, and 19.9% and 1.5% respectively in urban areas.[69]

The crime rate in the state in 2004 was 82.6 per 100,000, which was half of the national average.[93] This is the fourth-lowest crime rate among the 32 states and union territories of India.[94] However, the state reported the highest rate of Special and Local Laws (SLL) crimes.[95] In reported crimes against women, the state showed a crime rate of 7.1 compared to the national rate of 14.1.[94] Some estimates state that there are more than 60,000 brothel-based women and girls in prostitution in Kolkata.[96] The population of prostitutes in Sonagachi constitutes mainly of Nepalese, Indians and Bangladeshis.[96] Some sources estimate there are 60,000 women in the brothels of Kolkata.[96] The largest prostitution area in city is Sonagachi.[96] West Bengal was the first Indian state to constitute a Human Rights Commission of its own.[94]

Culture

Literature

Rabindranath Tagore is Asia's first Nobel laureate and composer of India's national anthem

Swami Vivekananda was a key figure in introducing Vedanta and Yoga in Europe and USA,[] raising interfaith awareness and making Hinduism a world religion.[]

The Bengali language boasts a rich literary heritage, shared with neighbouring Bangladesh. West Bengal has a long tradition in folk literature, evidenced by the *Charyapada, Mangalkavya, Shreekrishna Kirtana, Thakurmar Jhuli,* and stories related to Gopal Bhar. In the nineteenth and twentieth century, Bengali literature was modernized in the works of authors such as Bankim Chandra Chattopadhyay, Michael Madhusudan Dutt, Rabindranath Tagore, Kazi Nazrul Islam, Sharat Chandra Chattopadhyay, Jibananda Das and Manik Bandyopadhyay. In modern times Jibanananda Das, Bibhutibhushan Bandopadhyay, Tarashankar Bandopadhyay, Manik Bandopadhyay, Ashapurna Devi, Shirshendu Mukhopadhyay, Buddhadeb Guha, Mahashweta Devi, Samaresh Majumdar, Sanjeev Chattopadhyay and Sunil Gangopadhyay among others are well known.

Music and dance

Baul singers at Basanta-Utsab, Shantiniketan

The Baul tradition is a unique heritage of Bengali folk music, which has also been influenced by regional music traditions.[97] Other folk music forms include Gombhira and Bhawaiya. Folk music in West Bengal is often accompanied by the ektara, a one-stringed instrument. West Bengal also has an heritage in North Indian classical music. "Rabindrasangeet", songs composed and set into tune by Rabindranath Tagore and "Nazrul geeti" (by Kazi Nazrul Islam) are popular. Also prominent are other musical forms like Dwijendralal, Atulprasad and Rajanikanta's songs, and *"adhunik"* or modern music from films and other composers.

From the early 1990s, there has been an emergence and popularisation of new genres of music, including fusions of Baul and Jazz by several Bangla bands, as well as the emergence of what has been called *Jeebonmukhi Gaan* (a modern genre based on realism). Bengali dance forms draw from folk traditions, especially those of the tribal groups, as well as the broader Indian dance traditions. Chau dance of Purulia is a rare form of mask dance. State is known for Bengali folk music such as baul and kirtans and *gajan*, and modern songs including Bengali adhunik songs. From the early 1990s, there has been an emergence of new genres of music, including the emergence of what has been called

Dance with Rabindra Sangeet.

Bengali *Jeebonmukhi Gaan* (a modern genre based on realism) by artists like Anjan Dutta, Kabir Suman, Nachiketa and folk/alternative/rock bands like Moheener Ghoraguli, Chandrabindoo, Bhoomi, Cactus and Fossils. Dutta's songs are influenced by classical music, and especially country music and blues and Bob Dylan and Leonard Cohen which he fused with Bengali tradition of east west, as did Suman. American urban folk and grunge are also an inspiration for this generation.[98]

Films

Mainstream Hindi films are popular in Bengal, and the state is home to a thriving cinema industry, dubbed "Tollywood". Tollygunj in Kolkata is the location of numerous Bengali movie studios, and the name "Tollywood" (similar to Hollywood and Bollywood) is derived from that name. The Bengali film industry is well known for its art films, and has produced acclaimed directors like Satyajit Ray, Mrinal Sen, Tapan Sinha and Ritwik Ghatak. Prominent contemporary directors include Buddhadev Dasgupta, Goutam Ghose, Aparna Sen and Rituparno Ghosh.

Fine arts

Bengal had been the harbinger of modernism in fine arts. Abanindranath Tagore, called the father of Modern Indian Art had started the Bengal School of Art which was to create styles of art outside the European realist tradition which was taught in art colleges under the colonial administration of the British Government. The movement had many adherents like Gaganendranath Tagore, Ramkinkar Baij, Jamini Roy and Rabindranath Tagore. After Indian Independence, important groups like the Calcutta Group and the Society of Contemporary Artists were formed in Bengal which dominated the art scene in India.

Reformist heritage

The capital, Kolkata, was the workplace of several social reformers, like Raja Ram Mohan Ray, Iswar Chandra Vidyasagar, and Swami Vivekananda. These social reforms have eventually led to a cultural atmosphere where practices like sati, dowry, and caste-based discrimination or untouchability, the evils that crept into the Hindu society, were abolished.

Cuisine

Rice and fish are traditional favourite foods, leading to a saying in Bengali, *machhe bhate bangali*, that translates as "fish and rice make a Bengali".[99] Bengal's vast repertoire of fish-based dishes includes hilsa preparations, a favorite among Bengalis. There are numerous ways of cooking fish depending on the texture, size, fat content and the bones. Sweets occupy an important place in the diet of Bengalis and at their social ceremonies. It is an ancient custom among both Hindu and Muslim Bengalis to distribute sweets during festivities. The confectionery industry has flourished because of its close association with social and religious

Patisapta - A kind of Pitha; which is a popular sweet dish in West Bengal during winter.

ceremonies. Competition and changing tastes have helped to create many new sweets. Bengalis make distinctive sweetmeats from milk products, including *Rôshogolla*, *Chômchôm*, *Kalojam* and several kinds of *sondesh*. Pitha, a kind of sweet cake, bread or dimsum are specialties of winter season. Sweets like coconut-naru, til-naru, moa, payesh, etc. are prepared during the festival of Lakshmi puja. Popular street food includes *Aloor Chop*, Beguni, Kati roll, and phuchka.[100] [101]

The variety of fruits and vegetables that Bengal has to offer is incredible. A host of gourds, roots and tubers, leafy greens, succulent stalks, lemons and limes, green and purple eggplants, red onions, plantain, broad beans, okra, banana tree stems and flowers, green jackfruit and red pumpkins are to be found in the markets or anaj bazaar as popularly called. *Panta bhat* (rice soaked overnight in water)with onion & green chili is a traditional dish consumed in rural areas. Common spices found in a Bengali kitchen are cumin, ajmoda (radhuni), bay leaf, mustard, ginger, green chillies, turmeric, etc. People of erstwhile East Bengal use a lot of ajmoda, coriander leaves, tamarind, coconut and mustard in their cooking; while those aboriginally from West Bengal use a lot of sugar, garam masala and red chilli powder. Vegetarian dishes are mostly without onion and garlic.

A *Murti* (representation) of Maa Durga

Costumes

Bengali women commonly wear the *shaṛi* , often distinctly designed according to local cultural customs. In urban areas, many women and men wear Western attire. Among men, western dressing has greater acceptance. Men also wear traditional costumes such as the *panjabi* with *dhuti*, often on cultural occasions.

Festivals

Durga Puja in October is the most popular festival in the West Bengal.[102] Poila Baishakh (the Bengali New Year), Rathayatra, Dolyatra or Basanta-Utsab, Nobanno, *Poush Parbon* (festival of Poush), Kali Puja, SaraswatiPuja, LaxmiPuja, Christmas, Eid ul-Fitr, Eid ul-Adha and Muharram are other major festivals. Buddha Purnima, which marks the birth of Gautama Buddha, is one of the most important Hindu/Buddhist festivals while Christmas, called *Bôṛodin* (Great day) in Bengali is celebrated by the minority Christian population. Poush mela is a popular festival of Shantiniketan, taking place in winter. West Bengal has been home to several famous religious teachers, including Sri Chaitanya, Sri Ramakrishna, Swami Vivekananda, A. C. Bhaktivedanta Swami Prabhupada and Paramahansa Yogananda. The *swami* is credited with introducing Hinduism to western society and became a religious symbol of the nation in the eyes of the intellectuals of the west.

Education

West Bengal schools are run by the state government or by private organisations, including religious institutions. Instruction is mainly in English or Bengali, though Urdu is also used, especially in Central Kolkata. The secondary schools are affiliated with the Council for the Indian School Certificate Examinations (CISCE), the Central Board for Secondary Education (CBSE), the National Institute of Open School (NIOS) or the West Bengal Board of Secondary Education.[103] Under the 10+2+3 plan, after completing secondary school, students typically enroll for 2 years in a junior college, also known as pre-university, or in schools with a higher secondary facility affiliated with the West Bengal Council of Higher Secondary Education or any central board. Students choose from one of three streams, namely liberal arts, commerce or science. Upon completing the required coursework, students may enroll in general or professional degree programs.

Medical College Kolkata

St. Paul's School, Darjeeling (around 1905)

IIT Kharagpur

West Bengal has eighteen universities.[104] [105] The University of Calcutta, one of the oldest and largest public universities in India, has more than 200 affiliated colleges. Kolkata has played a pioneering role in the development of the modern education system in India. It is the gateway to the revolution of European education. Sir William Jones (philologist) established the Asiatic Society in 1794 for promoting oriental studies. People like Ram Mohan Roy, David Hare, Ishwar Chandra Vidyasagar and William Carey played a leading role in the setting up of modern schools and colleges in the city. The Fort William College was established in 1810. The Hindu College was established in 1817. In 1855 the Hindu College was renamed as the Presidency

College.[106] The Bengal Engineering & Science University and Jadavpur University are prestigious technical universities.[107] Visva-Bharati University at Santiniketan is a central university and an institution of national importance.[108] The state has several higher education institutes of national importance including Indian Institute of Foreign Trade, Indian Institute of Management Calcutta (the first IIM), Indian Institute of Science Education and Research, Kolkata, Indian Statistical Institute, Indian Institute of Technology Kharagpur (the first IIT), National Institute of Technology, Durgapur and West Bengal National University of Juridical Sciences. After 2003 the state govt supported the creation of West Bengal University of Technology, West Bengal State University and Gour Banga University.

Besides these, the state also has Kalyani University, The University of Burdwan, Vidyasagar University and North Bengal University-all well established and nationally renowned, to cover the educational needs at the district levels of the state and also an Indian Institute of Science Education and Research, Kolkata. Also recently Presidency College, Kolkata became a University named Presidency University. Apart from this there is another private university run by Ramakrishna mission named Ramakrishna Mission Vivekananda University at Belur Math. There are a number of research institutes in kolkata. The Indian Association for the Cultivation of Science is the first research institute

IIM Calcutta

in Asia. C. V. Raman got Nobel Prize for his discovery (Raman Effect) done in IACS. Also Bose Institute, Saha Institute of Nuclear Physics, S.N. Bose National Centre for Basic Sciences, Indian Institute of Chemical Biology, Central Glass and Ceramic Research Institute, Variable Energy Cyclotron Center are most prominent. A large number of Indian Scholars are educated at different universities in Bengal. State has produced likes of Jagadish Chandra Bose, Satyendra Nath Bose and RC Bose.

Media

West Bengal had 505 published newspapers in 2005,[109] of which 389 were in Bengali.[109] *Ananda Bazar Patrika*, published from Kolkata with 1,277,801 daily copies, has the largest circulation for a single-edition, regional language newspaper in India.[109] Other major Bengali newspapers are *Bartaman, Sangbad Pratidin, Aajkaal, Jago Bangla, Uttarbanga Sambad* and *Ganashakti*. Major English language newspapers which are published and sold in large numbers are *The Telegraph, The Times of India, Hindustan Times, The Hindu, The Statesman, The Indian Express* and *Asian Age*. Some prominent financial dailies like *The Economic Times, Financial Express, Business Line* and *Business Standard* are widely circulated. Vernacular newspapers such as those in Hindi, Nepali Gujarati, Oriya, Urdu and Punjabi are also read by a select readership.

Doordarshan is the state-owned television broadcaster. Multi system operators provide a mix of Bengali, Nepali, Hindi, English and international channels via cable. Bengali 24-hour television news channels include STAR Ananda, Tara Newz, Kolkata TV, News Time, 24 Ghanta, Mahuaa Khobor, Ne Bangla, CTVN Plus, Channel 10 and R Plus.[110] [111] All India Radio is a public radio station.[111] Private FM stations are available only in cities like Kolkata, Siliguri and Asansol.[111] Vodafone, Airtel, BSNL, Reliance Communications, Uninor, Aircel, MTS India, Tata Indicom, Idea Cellular and Tata DoCoMo are available cellular phone operators. Broadband internet is available in select towns and cities and is provided by the state-run BSNL and by other private companies. Dial-up access is provided throughout the state by BSNL and other providers.

Sports

Salt Lake Stadium – Yuva Bharati Krirangan, Kolkata

Cricket and football (soccer) are popular sports in the state. West Bengal, unlike most other states of India, is noted for its passion and patronage of football.[7] [8] [9] Kolkata is one of the major centers for football in India[112] and houses top national clubs such as East Bengal, Mohun Bagan and Mohammedan Sporting Club.[113] Indian sports such as Kho Kho and Kabaddi are also played. Calcutta Polo Club is considered as the oldest polo club of the world,[114] and the Royal Calcutta Golf Club is the oldest of its kind outside Great Britain.[115]

West Bengal has several large stadiums—The Eden Gardens is one of only two 100,000-seat cricket amphitheaters in the world, although renovations will reduce this figure.[116] Kolkata Knight Riders, East Zone and Bengal play there, and the 1987 World Cup final was there although in 2011 World Cup, Eden Gardens was stripped due to construction incompleteness. Salt Lake Stadium—a multi-use stadium—is the world's second highest-capacity football stadium.[117] [118] Calcutta Cricket and Football Club is the second-oldest cricket club in the world.[119] National and international sports events are also held in Durgapur, Siliguri and Kharagpur.[120] Notable sports persons

Eden Gardens in Kolkata

from West Bengal include former Indian national cricket captain Sourav Ganguly, Pankaj Roy Olympic tennis bronze medallist Leander Paes, and chess grand master Dibyendu Barua. Other major sporting icons over the years include famous football players such as Chuni Goswami, PK Banerjee and Sailen Manna as well as swimmer Mihir Sen and athlete Jyotirmoyee Sikdar (winner of gold medals at the Asian Games).[121]

See also

- Bengal Army
- Bengali calendar
- Bengali Language Movement
- History of India
- List of people from West Bengal

Notes

[1] "Area, population, decennial growth rate and density for 2001 and 2011 at a glance for West Bengal and the districts: provisional population totals paper 1 of 2011: West Bengal" (http://www.censusindia.gov.in/2011-prov-results/prov_data_products_wb.html). Registrar General & Census Commissioner, India. . Retrieved 26 January 2012.

[2] "Sex ratio, 0-6 age population, literates and literacy rate by sex for 2001 and 2011 at a glance for West Bengal and the districts: provisional population totals paper 1 of 2011: West Bengal" (http://www.censusindia.gov.in/2011-prov-results/prov_data_products_wb.html). Government of India:Ministry of Home Affairs. . Retrieved 29 January 2012.

[3] http://www.westbengal.gov.in/

[4] Special correspondent (19 August 2011). "West Bengal may be renamed PaschimBanga" (http://www.thehindu.com/news/national/article2373155.ece). *The Hindu* (Chennai, India). . Retrieved 7 February 2012.

[5] "India: Administrative Divisions (population and area)" (http://www.world-gazetteer.com/wg.php?x=&men=gadm&lng=en&des=wg&geo=-104&srt=npan&col=abcdefghinoq&msz=1500&va=x). Census of India. . Retrieved April 17, 2009.

[6] "Net state domestic product at factor cost—state-wise (at current prices)" (http://www.rbi.org.in/scripts/PublicationsView.aspx?id=13592). *Handbook of statistics on Indian economy.* Reserve Bank of India. 15 September 2011. . Retrieved 7 February 2012.

[7] Dineo, Paul; Mills, James (2001). *Soccer in South Asia: empire, nation, diaspora.* London: Frank Cass Publishers. p. 71. ISBN 978-0-7146-8170-2.

[8] Bose, Mihir (2006). *The magic of Indian cricket: cricket and society in India.* Psychology Press. p. 240. ISBN 9780415356916.

[9] Das Sharma, Amitabha (2002). "Football and the big fight in Kolkata" (http://www.la84foundation.org/SportsLibrary/FootballStudies/2002/FS0502g.pdf) (PDF). *Football Studies* **5** (2): 57. . Retrieved `5 April 2012.

[10] "Bangldesh: early history, 1000 B.C.–A.D. 1202" (http://memory.loc.gov/cgi-bin/query/r?frd/cstdy:@field(DOCID+bd0014)). *Bangladesh: A country study.* Washington, D.C.: Library of Congress. September 1988. . Retrieved 2 March 2012. "Historians believe that Bengal, the area comprising present-day Bangladesh and the Indian state of West Bengal, was settled in about 1000 B.C. by Dravidian-speaking peoples who were later known as the Bang. Their homeland bore various titles that reflected earlier tribal names, such as Vanga, Banga, Bangala, Bangal, and Bengal."

[11] Sarkar, Sebanti (March 28, 2008). "History of Bengal just got a lot older" (http://www.telegraphindia.com/1080328/jsp/frontpage/story_9067406.jsp). *The Telegraph* (Calcutta, India). . Retrieved 13 September 2010. "Humans walked on Bengal's soil 20,000 years ago, archaeologists have found out, pushing the state's pre-history back by some 8,000 years."

[12] Bharadwaj, G (2003). "The Ancient Period". In Majumdar, RC. *History of Bengal.* B.R. Publishing Corp.

[13] Sultana, Sabiha. "Settlement in Bengal (Early Period)" (http://www.banglapedia.org/httpdocs/HT/S_0221.HTM). *Banglapedia.* Asiatic Society of Bangladesh. . Retrieved 4 March 2012.

[14] Chowdhury, AM. "Gangaridai" (http://banglapedia.search.com.bd/HT/G_0019.htm). *Banglapedia.* Asiatic Society of Bangladesh. . Retrieved 8 September 2006.

[15] Prasad, Prakash Chandra (2003). *Foreign trade and commerce in ancient India* (http://books.google.com/?id=mFW3sXnzEQ4C&pg=PA231&dq=ancient+history+of+bengal+trade#v=onepage&q=bengal&f=false). New Delhi: Abhinav Publications. p. 28. ISBN 978-81-7017-053-2. . Retrieved 13 September 2010.

[16] Geiger, Wilhelm (2003) [1908]. "Chapter VI: The Coming of Viajaya" (http://lakdiva.org/mahavamsa/chap006.html). *Mahavamsa: Great Chronicle of Ceylon* (http://books.google.com/?id=nX2af3kcregC&printsec=frontcover&dq=wilhelm+geiger#v=onepage&q&f=false). New Delhi: Asian Educational Services. pp. 51–54. ISBN 81-206-0218-8. . Retrieved 2 March 2012.

[17] Bhattacharyya, P.K.. "Shashanka" (http://www.banglapedia.org/httpdocs/HT/S_0122.HTM). *Banglapedia.* Asiatic Society of Bangladesh. . Retrieved 2 March 2012.

[18] Chaudhury, S; Mohsin, KM. "Sirajuddaula" (http://www.banglapedia.org/httpdocs/HT/S_0411.HTM). *Banglapedia.* Asiatic Society of Bangladesh. . Retrieved 2 March 2012.

[19] Fiske, John. "The famine of 1770 in Bengal" (http://etext.library.adelaide.edu.au/f/fiske/john/f54u/chapter9.html). *The Unseen World, and other essays.* Adelaide: University of Adelaide Library Electronic Texts Collection. . Retrieved 26 October 2006.

[20] (Baxter 1997, pp. 30–32)

[21] (Baxter 1997, pp. 39–40)

[22] Wolpert, Stanley (1999). *India* (http://books.google.com/books?id=nHnOERqf-MQC). Berkeley, California, USA: University of California Press. p. 14. ISBN 978-0-520-22172-7. . Retrieved 2 March 2012.

[23] Islam, Sirajul. "Partition of Bengal, 1947" (http://www.banglapedia.org/httpdocs/HT/P_0101.HTM). *Banglapedia.* Asiatic Society of Bangladesh. . Retrieved 3 March 2012.

[24] Dr. Sailen Debnath, 'West Bengal in Doldrums'ISBN 978-81-86860-34-2; & Dr. Sailen Debnath ed. Social and Political Tensions In North Bengal since 1947, ISBN 81-86860-23-1

[25] Dr. Sailen Debnath,ed. Social and Political Tensions In North Bengal since 1947, ISBN 81-86860-23-1.

[26] (Bennett & Hindle 1996, pp. 63–70)

[27] Biswas, Soutik (16 April 2006). "Calcutta's colourless campaign" (http://news.bbc.co.uk/2/hi/south_asia/4909832.stm). BBC. . Retrieved 15 February 2012.

[28] Ghosh Roy, Paramasish (22 July 2005). "Maoist on rise in West Bengal" (http://www.voanews.com/bangla/archive/2005-07/2005-07-22-voa10.cfm). *VOA Bangla.* Voice of America. . Retrieved 11 September 2006.

[29] "Maoist Communist Centre (MCC)" (http://www.satp.org/satporgtp/countries/india/terroristoutfits/MCC.htm). *Left-wing extremist group*. South Asia Terrorism Portal. . Retrieved 11 September 2006.

[30] "Several hurt in Singur clash" (http://www.rediff.com/news/2007/jan/28singur.htm). *rediff news*. 28 January 2007. . Retrieved 15 March 2007.

[31] "Red-hand Buddha: 14 killed in Nandigram re-entry bid" (http://www.telegraphindia.com/1070315/asp/frontpage/story_7519166.asp). *The Telegraph* (Calcutta, India). 15 March 2007. . Retrieved 15 March 2007.

[32] "8 Indian states have more poor than 26 poorest African nations" (http://timesofindia.indiatimes.com/India/8-Indian-states-have-more-poor-than-26-poorest-African-nations/articleshow/6158960.cms). *Times of India* (New Delhi). 22 July 2010. . Retrieved 4 March 2012.

[33] "WB takes the cake when it comes to bandhs" (http://economictimes.indiatimes.com/News/PoliticsNatio/WB_takes_the_cake_when_it_comes_to_bandhs/articleshow/842652.cms). *Economic Times* (New Delhi). 18 December 2006. . Retrieved 4 March 2012.

[34] "Business in West Bengal affected ahead of Tuesday strike" (http://sify.com/finance/business-in-west-bengal-affected-ahead-of-tuesday-strike-news-default-ke0sabdcibh.html). *sify finance*. 26 April 2010. . Retrieved 4 March 2012.

[35] Roy, Hiranmoy; Bhattacharjee, Kaushik (August 2009). "Convergence of human development across Indian States" (http://www.igidr.ac.in/pdf/publication/PP-062-22.pdf) (PDF). Indira Gandhi Institute of Development Research. p. 4. . Retrieved 4 March 2012.

[36] Shah, Mansi (2007). "Waiting for health care: a survey of a public hospital in Kolkata" (http://ccs.in/ccsindia/downloads/intern-papers-08/Waiting-for-Healthcare-A-survey-of-a-public-hospital-in-Kolkata-Mansi.pdf) (PDF). Centre for Civil Society. . Retrieved 31 January 2012.

[37] "West Bengal: health systems development initiative programme memorandum" (http://www.wbhealth.gov.in/Externally_Aided_Projects/HSDI-DFID Programme Memorandum.pdf) (PDF). Government of West Bengal. 15 January 2005. . Retrieved 4 March 2012.

[38] "Impact of social sector development in West Bengal – Midnapore and Birbhum districts" (http://planningcommission.nic.in/reports/sereport/ser/wbm_indx.htm). Planning Commission of India. . Retrieved 4 March 2012.

[39] Mukherjee, Rudrangshu (5 October 2008). "Murder, most foul - the people of Bengal created the darkness that envelops them" (http://www.telegraphindia.com/1081005/jsp/opinion/story_9927371.jsp). *The Telegraph* (Kolkata). . Retrieved 4 March 2012.

[40] "ADB pep pill for Bengal" (http://www.telegraphindia.com/1100613/jsp/business/story_12560050.jsp). *The Telegraph* (Kolkata). 13 June 2010. . Retrieved 4 March 2012.

[41] Ramesh, Randeep (12 November 2007). "Six killed as farmers and communists clash in West Bengal" (http://www.guardian.co.uk/world/2007/nov/12/india.randeepramesh). *The Guardian* (London). . Retrieved 4 March 2012.

[42] "West Bengal political violence continues" (http://economictimes.indiatimes.com/news/politics/nation/West-Bengal-political-violence-continues/articleshow/4871906.cms). *Economic Times* (New Delhi). 8 August 2009. . Retrieved 4 March 2012.

[43] Pal, Supratim (14 May 2007). "Top of world in kingdom of cloud" (http://www.telegraphindia.com/1070514/asp/ranchi/story_7772890.asp). *The Telegraph* (Kolkata). . Retrieved 16 February 2012.

[44] Jayapalan, N (2001). *Foreign policy of India*. New Delhi: Atlantic Publishers and Distributors. p. 344. ISBN 81-7156-898-X.

[45] "Alarming rise in bacterial percentage in Ganga waters" (http://www.thehindubusinessline.in/2006/08/04/stories/2006080402921900.htm). *The Hindu Business Line* (Chennai). 4 August 2006. . Retrieved 4 March 2012.

[46] "Groundwater Arsenic Contamination Status in West Bengal" (http://www.soesju.org/arsenic/wb.htm). *Groundwater Arsenic Contamination in West Bengal – India (17 Years Study)*. School of Environmental Studies, Jadavpur University. . Retrieved October 29, 2006.

[47] "Climate" (http://www.webindia123.com/westbengal/land/climate.htm). *West Bengal: Land*. Suni System (P) Ltd. . Retrieved September 5, 2006.

[48] "kal Baisakhi" (http://amsglossary.allenpress.com/glossary/search?id=kal-baisakhi1). *Glossary of Meteorology*. American Meteorological Society. . Retrieved September 5, 2006.

[49] "State animals, birds, trees and flowers" (http://web.archive.org/web/20090304232302/http://www.wii.gov.in/nwdc/state_animals_tree_flowers.pdf) (PDF). Wildlife Institute of India. Archived from the original (http://www.wii.gov.in/nwdc/state_animals_tree_flowers.pdf) on 4 March 2009. . Retrieved 5 March 2012.

[50] "Forest and tree resources in states and union territories: West Bengal" (http://www.fsi.nic.in/sfr_2009/westbengal.pdf) (PDF). *India state of forest report 2009*. Forest Survey of India, Ministry of Environment & Forests, Government of India. pp. 163–166. . Retrieved 4 March 2012.

[51] "Forest cover" (http://www.fsi.nic.in/sfr_2009/chapter2.pdf) (PDF). *India state of forest report 2009*. Forest Survey of India, Ministry of Environment & Forests, Government of India. pp. 14–24. . Retrieved 4 March 2012.

[52] "Environmental Issues" (http://hdr.undp.org/en/reports/national/asiathepacific/india/India_West Bengal_2004_en.pdf) (PDF). *West Bengal Human Development Report 2004*. Development and Planning Department, Government of West Bengal. May 2004. pp. 180–182. ISBN 81-7955-030-3. . Retrieved 5 March 2012.

[53] Islam, Sadiq (29 June 2001). "World's largest mangrove forest under threat" (http://archives.cnn.com/2001/fyi/student.bureau/06/29/sundarbans/index.html). CNN. . Retrieved 31 October 2006.

[54] Mukherji, S.J. (2000). *College Botany Vol. III: (chapter on Phytogeography)*. Calcutta: New Central Book Agency. pp. 345–365.

[55] "Sundarbans National Park" (http://whc.unesco.org/en/list/452). *World heritage list.* UNESCO World Heritage Center. . Retrieved 4 March 2012.

[56] "Natural vegetation" (http://www.webindia123.com/westbengal/land/forest.htm#N). *West Bengal.* Suni System (P) Ltd. . Retrieved 31 October 2006.

[57] "West Bengal: General Information" (http://web.archive.org/web/20060819094729/http://www.indiainbusiness.nic.in/indian-states/westbengal/General.htm). *India in Business.* Federation of Indian Chambers of Commerce and Industry. Archived from the original (http://www.indiainbusiness.nic.in/indian-states/westbengal/General.htm) on 19 August 2006. . Retrieved 25 August 2006.

[58] "Problems of Specific Regions" (http://hdr.undp.org/en/reports/national/asiathepacific/india/India_West Bengal_2004_en.pdf) (PDF). *West Bengal Human Development Report 2004.* Development and Planning Department, Government of West Bengal. May 2004. pp. 200–203. ISBN 81-7955-030-3. . Retrieved 5 March 2012.

[59] "West Bengal legislative assembly" (http://legislativebodiesinindia.gov.in/States\westbengal\wesbengal-w.htm). *Legislative bodies in India.* National Informatics Centre, India. . Retrieved October 28, 2006.

[60] Delimitation Commission (15 February 2006). "Notification: order no. 18" (http://ceowestbengal.nic.in/news_pdf/gazette123.pdf) (PDF). New Delhi: Election Commission of India. pp. 23–25. . Retrieved 11 February 2012.

[61] "Composition of Rajya Sabha" (http://rajyasabha.nic.in/rsnew/rsat_work/chapter-2.pdf) (PDF). *Rajya Sabha at work.* New Delhi: Rajya Sabha Secretariat. pp. 24–25. . Retrieved 15 February 2012.

[62] "Statewise results - West Bengal" (http://eciresults.nic.in/statewiseS25.htm). Election Commission of India. . Retrieved 13 May 2011.

[63] "Directory of district, sub division, panchayat samiti/ block and gram panchayats in West Bengal, March 2008" (http://www.webel-india.com/blocks n grampanchayats.doc) (DOC). West Bengal Electronics Industry Development Corporation Limited, Government of West Bengal. March 2008. p. 1. . Retrieved 15 February 2012.

[64] "Section 2 of West Bengal Panchayat Act, 1973" (http://wbdemo5.nic.in/html/asp/g2csw/sections/2.htm). Department of Panchayat and Rural Department, West Bengal. . Retrieved December 9, 2008.

[65] "Urban agglomerations/cities having population 1 million and above" (http://censusindia.gov.in/2011-prov-results/paper2/data_files/india2/Million_Plus_UAs_Cities_2011.pdf). *Provisional population totals, census of India 2011.* The Registrar General & Census Commissioner, India. 2011. . Retrieved 26 January 2012.

[66] "Cities having population 1 lakh and above, census 2011" (http://www.censusindia.gov.in/2011-prov-results/paper2/data_files/India2/Table_2_PR_Cities_1Lakh_and_Above.pdf). *Provisional population totals, census of India 2011.* The Registrar General & Census Commissioner, India. . Retrieved 18 October 2011.

[67] "West Bengal" (http://www.ibef.org/download/West_Bengal_271211.pdf). India Brand Equity Foundation. November 2011. . Retrieved 6 February 2012.

[68] "Industrial infrastructure" (http://www.wbidc.com/about_wb/industrial_infrastructure.htm). West bengal Industrial Development Corporation. . Retrieved 5 March 2012.

[69] "Introduction and Human Development Indices for West Bengal" (http://hdr.undp.org/en/reports/national/asiathepacific/india/India_West Bengal_2004_en.pdf) (PDF). *West Bengal Human Development Report 2004.* Development and Planning Department, Government of West Bengal. May 2004. pp. 4–6. ISBN 81-7955-030-3. . Retrieved 5 March 2012.

[70] "Mamata seeks debt restructuring plan for West Bengal" (http://articles.economictimes.indiatimes.com/2011-10-22/news/30309832_1_debt-restructuring-plan-debt-burden-12th-plan). *Economic Times* (New Delhi). 22 October 2011. . Retrieved 4 March 2012.

[71] Ray Choudhury, Ranabir (27 October 2006). "A new dawn beckons West Bengal" (http://www.thehindubusinessline.com/2006/10/27/stories/2006102700080100.htm). *The Hindu Business Line* (Chennai). . Retrieved 29 October 2006.

[72] "West Bengal Industrial Development Corporation Ltd." (http://www.indiaathannover.org/pdf/exhibitorslist.pdf) (PDF). *India @ Hannover Messe 2006.* Engineering Export Promotion Council (EEPC), India. p. 303. . Retrieved September 7, 2006.

[73] Saha, Sambit (9 September 2003). "Nathula trade may spur business in NE" (http://www.rediff.com/money/2003/sep/09trading.htm). rediff.com. . Retrieved 18 September 2007.

[74] Raja Mohan, C. (16 July 2004). "A foreign policy for the East" (http://www.hindu.com/2004/07/16/stories/2004071601841000.htm). Chennai: The Hindu. . Retrieved 5 March 2012.

[75] "Statewise Length of national highways in India" (http://morth.nic.in/showfile.asp?lid=366). *National Highways.* Department of Road Transport and Highways; Ministry of Shipping, Road Transport and Highways; Government of India. . Retrieved 9 February 2012.

[76] Chattopadhyay, Suhrid Sankar (January–February 2006). "Remarkable Growth" (http://www.flonnet.com/fl2302/stories/20060210004209800.htm). *Frontline* (Chennai, India: The Hindu) **23** (2). . Retrieved March 31, 2008.

[77] "Geography : Railway Zones" (http://www.irfca.org/faq/faq-geog.html). *IRFCA.org.* Indian Railways Fan Club. . Retrieved August 31, 2007.

[78] "About Kolkata Metro" (http://www.kolmetro.com/). Kolkata Metro. . Retrieved September 1, 2007.

[79] "Mountain Railways of India" (http://whc.unesco.org/en/list/944). UNESCO World Heritage Centre. . Retrieved April 30, 2006.

[80] "Port info: cargo statistics" (http://www.kolkataporttrust.gov.in/). *Kolkata Port Trust.* Kolkata Port Trust, India. . Retrieved 9 February 2012.

[81] "Intra-city train travel" (http://timesfoundation.indiatimes.com/articleshow/657741.cms). *reaching India* (Times Internet Limited). . Retrieved August 31, 2007.

[82] "India ferry disaster kills scores" (http://www.bbc.co.uk/news/world-south-asia-11679012). *BBC News.* November 2, 2010. .

[83] "Census Population" (http://indiabudget.nic.in/es2006-07/chapt2007/tab97.pdf) (PDF). *Census of India*. Ministry of Finance India. .
     Retrieved December 18, 2008.

[84] Hoddie, Matthew (2006). *Ethnic realignments: a comparative study of government influences on identity* (http://books.google.com/
     books?id=6ka0nMJgKbYC). Lexington Books. pp. 114–115. ISBN 978-0-7391-1325-7. . Retrieved 16 February 2012.

[85] Banerjee, Himadri; Gupta, Nilanjana; Mukherjee, Sipra, eds. (2009). *Calcutta mosaic: essays and interviews on the minority communities of
     Calcutta* (http://books.google.com/books?id=cSTEOx_Lw9MC&dq). Anthem Press. p. 3. ISBN 978-81-905835-5-8. . Retrieved 29
     January 2012.

[86] Banerjee, Himadri; Gupta, Nilanjana; Mukherjee, Sipra, eds. (2009). *Calcutta mosaic: essays and interviews on the minority communities of
     Calcutta* (http://books.google.com/books?id=cSTEOx_Lw9MC&dq). Anthem Press. pp. 9–10. ISBN 978-81-905835-5-8. . Retrieved 29
     January 2012.

[87] "Data on Religion" (http://web.archive.org/web/20070812142520/http://www.censusindia.net/religiondata/). *Census of India (2001)*.
     Office of the Registrar General & Census Commissioner, India. Archived from the original (http://www.censusindia.gov.in/) on August
     12, 2007. . Retrieved August 26, 2006.

[88] "Report of the Commissioner for linguistic minorities: 47th report (July 2008 to June 2010)" (http://nclm.nic.in/shared/linkimages/
     NCLM47thReport.pdf). Commissioner for Linguistic Minorities, Ministry of Minority Affairs, Government of India. pp. 122–126. .
     Retrieved 16 February 2012.

[89] Population of West Bengal (80,221,171) is 7.8% of India's population (1,027,015,247)

[90] "Table 1: Distribution of population, sex ratio, density and decadal growth rate of population: 2011" (http://www.censusindia.gov.in/
     2011-prov-results/prov_results_paper1_india.html). *Provisional population totals paper 1 of 2011 India: series 1*. Registrar General &
     Census Commissioner, India. . Retrieved 16 February 2012.

[91] "Table 2(3): Literates and literacy rates by sex : 2011" (http://www.censusindia.gov.in/2011-prov-results/prov_results_paper1_india.
     html). *Provisional population totals paper 1 of 2011 India: series 1*. Registrar General & Census Commissioner, India. . Retrieved 16
     February 2012.

[92] "An Indian life: Life expectancy in our nation" (http://www.indiatogether.org/health/infofiles/life.htm). *India Together*. Civil Society
     Information Exchange Pvt. Ltd. . Retrieved August 26, 2006.

[93] National Crime Records Bureau (2004). "Crimes in Mega Cities" (http://ncrb.nic.in/crime2004/cii-2004/CHAP2.pdf) (PDF). *Crime in
     India-2004*. Ministry of Home Affairs. p. 158. . Retrieved August 26, 2006.

[94] "Human security" (http://hdr.undp.org/en/reports/national/asiathepacific/india/India_West Bengal_2004_en.pdf) (PDF). *West Bengal
     Human Development Report 2004*. Development and Planning Department, Government of West Bengal. May 2004. pp. 167–172.
     ISBN 81-7955-030-3. . Retrieved 5 March 2012.

[95] National Crime Records Bureau (2004). "General Crime Statistics Snapshots 2004" (http://ncrb.nic.in/crime2004/cii-2004/Snapshots.
     pdf) (PDF). *Crime in India-2004*. Ministry of Home Affairs. p. 1. . Retrieved April 26, 2006.

[96] "Plight of prostitutes in Kolkata" (http://www.merinews.com/article/plight-of-prostitutes-in-kolkata/137536.shtml). Merinews.com. .
     Retrieved 2012-01-09.

[97] "The Bauls of Bengal" (http://bengalonline.sitemarvel.com/bengali-folklore.asp?art=baul). *Folk Music*. BengalOnline. . Retrieved
     October 26, 2006.

[98] "Chau: The Rare Mask Dances" (http://www.boloji.com/dances/00109.htm). *Dances of India*. Boloji.com. . Retrieved October 22, 2006.

[99] Gertjan de Graaf, Abdul Latif. "Development of freshwater fish farming and poverty alleviation: A case study from Bangladesh" (http://
     govdocs.aquake.org/cgi/reprint/2003/1201/12010300.pdf) (PDF). Aqua KE Government. . Retrieved October 22, 2006.

[100] Saha, S (January 18, 2006). "Resurrected, the kathi roll – Face-off resolved, Nizam's set to open with food court" (http://www.
     telegraphindia.com/1060118/asp/calcutta/story_5733258.asp). Calcutta, India: The Telegraph (Kolkata). . Retrieved October 26, 2006.

[101] "Mobile food stalls" (http://www.bangalinet.com/mobile_foodstalls.htm). Bangalinet.com. . Retrieved October 26, 2006.

[102] "Durga Puja" (http://www.westbengaltourism.gov.in/web/guest/festival-home). *Festivals celebrated throughout West Bengal*.
     Department of Tourism, Government of West Bengal. . Retrieved 5 March 2012.

[103] "Boards of secondary & senior secondary education in India" (http://mhrd.gov.in/recognized_boards). Department of School Education
     and Literacy, Ministry of Human Resource Development, Government of India. . Retrieved 18 April 2012.

[104] "UGC recognised Universities in West Bengal with NAAC accreditation status" (http://www.educationobserver.com/resources/
     universsities/west_bengal.htm). Education Observer. . Retrieved October 26, 2006.

[105] "West Bengal University of Health Sciences" (http://www.thewbuhs.org/). West Bengal University of Health Sciences. . Retrieved
     October 26, 2006.

[106] "List of Affiliated Colleges" (http://web.archive.org/web/20080201164051/http://www.caluniv.ac.in/coll.htm). University of
     Calcutta. Archived from the original (http://www.caluniv.ac.in/coll.htm) on February 1, 2008. . Retrieved March 29, 2008.

[107] Mitra, P (August 31, 2005). "Waning interest" (http://www.telegraphindia.com/1050831/asp/careergraph/story_5174502.asp).
     Careergraph (Calcutta, India: The Telegraph). . Retrieved October 26, 2006.

[108] "Visva-Bharati: Facts and Figures at a Glance" (http://www.visva-bharati.ac.in/at_a_glance/at_a_glance.htm). Visva-Bharati
     Computer Centre. . Retrieved March 31, 2007.

[109] "General Review" (https://rni.nic.in/pii.asp). Registrar of Newspapers for India. . Retrieved 6 March 2012.

[110] "Bengali News Channel took 5 months to reach no.1 position" (http://www.moneycontrol.com/news/business/
     bengali-news-channel-took-5-months-to-reach-no1-position_242437.html). News Center. . Retrieved Sep 7, 2006.

[111] "CALCUTTA : Television, Radio Channels" (http://www.calcuttaweb.com/tvradio.shtml). Calcutta Web. . Retrieved Sep 7, 2006.

[112] Prabhakaran, Shaji (January 18, 2003). "Football in India – A Fact File" (http://www.longlivesoccer.com/indiafootball.htm). LongLiveSoccer.com. . Retrieved October 26, 2006.

[113] "Indian Football Clubs" (http://www.iloveindia.com/sports/football/clubs/index.html). Iloveindia.com. . Retrieved October 26, 2006.

[114] "History of Polo" (http://www.hpa-polo.co.uk/about/history_polo.asp). Hurlingham Polo Association. . Retrieved August 30, 2007.

[115] "Royal Calcutta Golf Club" (http://www.britannica.com/eb/topic-511285/Royal-Calcutta-Golf-Club). Encyclopaedia Britannica. . Retrieved August 30, 2007.

[116] "India – Eden Gardens (Kolkata)" (http://www.cricketweb.net/country/venue.php?CategoryIDAuto=12&VenueIDAuto=26). Cricket Web. . Retrieved October 26, 2006.

[117] "100 000+ Stadiums" (http://www.worldstadiums.com/stadium_menu/stadium_list/100000.shtml). World Stadiums. . Retrieved October 26, 2006.

[118] "The Asian Football Stadiums (30.000+ capacity)" (http://www.fussballtempel.net/afc/listeafc.html). Gunther Lades. . Retrieved October 26, 2006.

[119] Raju, Mukherji (March 14, 2005). "Seven Years? Head Start" (http://www.telegraphindia.com/1050314/asp/opinion/story_4428341. asp). Calcutta, India: The Telegraph. . Retrieved October 26, 2006.

[120] "Sports & Adventure" (http://www.wbtourism.com/sports_adventure/index.htm). West Bengal Tourism. . Retrieved October 22, 2006.

[121] "Famous Indian Football Players" (http://www.iloveindia.com/sports/football/players/index.html). Iloveindia.com. . Retrieved October 26, 2006.

References

- Baxter, C (1997). *Bangladesh, From a Nation to a State*. Westview Press. p. 0813336325. ISBN 1-85984-121-X
- Bennett, A; Hindle, J (1996). *London Review of Books: An Anthology*. Verso. pp. 63–70. ISBN 1-85984-121-X
- Roy, A; Alsayyad, N (2004). *Urban Informality: Transnational Perspectives from the Middle East, Latin America and South Asia*. Lexington Books. ISBN 0-7391-0741-0
- *West Bengal Human Development Report 2004* (http://hdr.undp.org/en/reports/national/asiathepacific/india/ India_West Bengal_2004_en.pdf). Development and Planning Department, Government of West Bengal. 2004. ISBN 81-7955-030-3
- *Impact of Social Sector Development in West Bengal* (http://planningcommission.nic.in/reports/sereport/ser/ wbm_indx.htm). Planning Commission, Government of India. 2009.
- Klass, L; Morton, S (1996). *Community Structure and industrialization in West Bengal*. University Press of America Inc.. ISBN 0-7618-0420-X
- Sunny, C (1999). "Poverty and social development in west bengal" (http://planningcommission.nic.in/reports/ sereport/ser/wbm/wbm_ch2.pdf). *India Rural Development Report, NIRD*. Retrieved 1999
- KPMG India, V (December 10, 2001). "Sustainable economic development in West Bengal – A Perspective" (http://www.in.kpmg.com/TL_Files/Pictures/West_Bengal.pdf). *Confederation of Indian Industry (CII)*. Retrieved 2007
- Amrita Basu, V. (1997). *Two Faces of Protest: Contrasting Modes of Women's Activism in India* (http://books. google.com/?id=ZyY0Yb5BrqgC&pg=PA25&dq=communism+in+west+bengal&cd=5#v=onepage& q=communism in west bengal). University of California Press ltd.. ISBN 0-520-06506-9. Retrieved June 16, 2009.
- Jasodhara Bagchi, Sarmistha Dutta Gupta, V. (2000). *The changing status of women in West Bengal, 1970–2000: the challenge ahead* (http://books.google.com/?id=KYYW8Un5zFAC&pg=PA119&dq=violence+west+ bengal&cd=1#v=onepage&q=violence west bengal). Saga Publication India Pvt Ltd.. ISBN 0-7619-3242-9. Retrieved June 16, 2010.
- Magnus Öberg, Kaare Strom, V. (2008). *Resources, governance and civil conflict* (http://books.google.com/ ?id=eBW-KtJ28ZsC&pg=PA93&dq=Naxalite+in+west+bengal&cd=1#v=onepage&q=Naxalite in west bengal). Routledge. ISBN 978-0-415-41671-9. Retrieved June 16, 2004.
- Atul Kohli, I. (1987). *The State and Poverty in India* (http://books.google.com/?id=vxLAK8EXo84C& pg=PA117&dq=poverty+in+west+bengal&cd=1#v=onepage&q= west bengal). Cambridge University Press. ISBN 978-0-521-37876-5. Retrieved June 16, 2007.

- Marvin, Davis (1983) [1983]. *Rank and rivalry: the politics of inequality in rural West Bengal.* Cambridge: Cambridge University Press. ISBN 0-521-24657-1.
- Richard Maxwell Eaton, The rise of Islam and the Bengal frontier, 1204–1760, 1993, University of California Press, California, California,1993, ISBN 0-520-08077-7.
- Ross Mallick. (1955). Development Policy of a Communist Government: West Bengal Since 1977, Cambridge University Press, Cambridge (Reprinted 2008) ISBN 978-0-521-43292-4.
- Jasodhara Bagchi, Sarmistha Dutta Gupta, V. (2009). *A Story of Ambivalent Modernization in Bangladesh and West Bengal: The Rise and Fall of Bengali Elitism in South Asia.* Peter Lang Publishing; First printing edition. ISBN 978-1-4331-0820-4.
- Tapan Raychaudhuri (2002). *Europe Reconsidered: Perceptions of the West in Nineteenth-Century Bengal.* Oxford University Press. ISBN 978-0-19-566109-5.
- Harriss-White, Barbara (editor) (2008). *Rural Commercial Capital: Agricultural Markets in West Bengal.* Oxford University Press, USA. ISBN 0-19-569159-8.
- Raychaudhuri, Ajitava (editor); Das, Tuhin K. (editor) (2005). *West Bengal economy: some contemporary issues* (http://books.google.com/?id=NTeHPuhTsXcC&pg=PA45&dq=politics+in+west+bengal& cd=42#v=onepage&q). Jadavpur University Press, India. ISBN 81-7764-731-8.
- Inden; Ronald B.; Ralph W (2005). *Kinship in Bengali Culture.* The University of Chicago Press, 1977. ISBN 81-8028-018-7.
- Davis, Marvin (1983). *Rank and rivalry: the politics of inequality in rural West Bengal.* 1st edition. Cambridge University Press. xxvii, 239. ISBN 0-521-24657-1.
- Banerjee, Anuradha (1998). *Environment, population, and human settlements of Sundarban Delta.* Ashok Kumar Mittal. ISBN 81-7022-739-9.

External links

Government

- Official West Bengal Government Web Portal (http://www.westbengal.gov.in/)
- Department of Tourism, Government of West Bengal (http://www.westbengaltourism.gov.in/web/guest/index)
- Directorate of Commercial Taxes, Government of West Bengal (http://www.wbcomtax.nic.in/welcome.asp)
- West Bengal Information Commission (http://wbic.gov.in/)

Other

- West Bengal travel guide from Wikitravel
- West Bengal (http://www.britannica.com/EBchecked/topic/640088/West-Bengal) *Encyclopædia Britannica* entry
- West Bengal (http://www.dmoz.org/Regional/Asia/India/West_Bengal/) at the Open Directory Project
- West Bengal (http://www.wikimapia.org/#lat=22.5697&lon=88.3697&z=10&l=0&m=b) Satellite view at WikiMapia

Gram_panchayat

Gram panchayats are local self-governments at the village or small town level in India. As of 2002 there were about 265,000 gram panchayats in India. The gram panchayat is the foundation of the Panchayat System. A gram panchayat can be set up in villages with minimum population of 300. Sometimes two or more villages are clubbed together to form group-gram panchayat when the population of the individual villages is less than 300.

Indian municipal organisation by state

Structure

The Sarpanch or Chairperson is the head of the Gram Panchayat. The elected members of the Gram Panchayat elect from among themselves a Sarpanch and a Deputy Sarpanch for a term of five years. In some places the panchayat president is directly elected by village people. The Sarpanch presides over the meetings of the Gram Panchayat and supervises its working. He implements the development schemes of the village. The Deputy Sarpanch, who has the power to make his own decisions, assists the Sarpanch in his work.

The Sarpanch has the responsibilities of

1. Looking after street lights, construction and repair work of the roads in the villages and also the village markets, fairs,collection of tax, festivals and celebrations.
2. Keeping a record of births, deaths and marriages in the village.
3. Looking after public health and hygiene by providing facilities for sanitation and drinking water.
4. Providing for education.
5. To organize the meetings of Gramsabha(ग्रामसभा) and Grampanchayat(ग्रामप ंचायत).

A grampannchyat consists of a between 7 and 17 members, elected from the wards of the village, and they are called a "panch". Peoples of village selects panch, with one-eighth of seats reserved for female candidates. To establish a Grampanchyat in a village, the population of the village should be at least 500 people of voting age.

Sources of income

The main source of income of the Gram Panchayat is the property tax levied on the buildings and the open spaces within the village. Other sources of income include professional tax, taxes on pilgrimage, animal trade, grant received from the State Government in proportion of land revenue and the grants received from the Zilla Parishad.

The gramsevak / gram vikas adhikari is communicator in government and village panchayat and do works for sarpanch. This post is in maharashtra, india. he is responsible person as sarpanch.

Principles of decentralisation

Dr S B Sen committee, a committee appointed by the Government of Kerala in 1996, had suggested the following principles, which was later adopted by the Second Administrative Reforms Commission, for local governance :-

- subsidiarity
- democratic decentralisation
- delineation of functions
- devolution of functions in real terms
- convergence
- citizen centricity

gram sabha is conducted six times in a year ...april, 1st may,15TH augest, 2octo, nov, 26 jan.

References

External links

- *Our Civic Life (Civics and Administration).* Maharashtra State Bureau of Textbook Production and Curriculum Research, Pune
- Subramaniam Vincent (2002-02-28). "Ugly duckling to swan" (http://www.indiatogether.org/2003/apr/ gov-kardcrefm.htm). India Together.

Community_development_block_in_India

The **Community development block** (Hindi: सामुदायकि वकिास ख ·ड) is a rural area earmarked for administration and development in India. The area is administered by a Block Development Officer. A community development block covers several gram panchayats, local administrative unit at the village level.

History

The community development programme was launched on a pilot basis in 1952 to provide for a substantial increase in the country's agricultural programme, and for improvements in systems of communication, in rural health and hygiene, and in rural education. The community development programme was rapidly implemented. In 1956, by the end of the first five year plan period, there were 248 blocks, covering around a fifth of the population in the country. By the end the second five year plan period, there were 3,000 blocks covering 70 per cent of the rural population. By 1964, the entire country was covered.[1]

References

[1] "The Failure of the Community Development Programme in India" (http://cdj.oxfordjournals.org/cgi/pdf_extract/11/2/95). . Retrieved 2010-04-06.

Bhangri_Pratham_Khanda

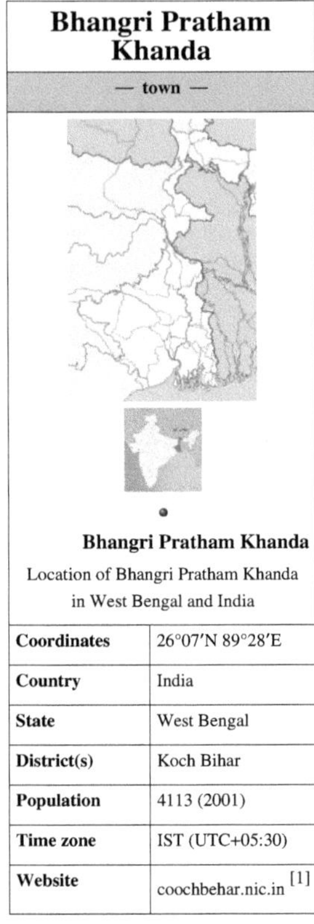

Location of Bhangri Pratham Khanda
in West Bengal and India

Coordinates	26°07′N 89°28′E
Country	India
State	West Bengal
District(s)	Koch Bihar
Population	4113 (2001)
Time zone	IST (UTC+05:30)
Website	coochbehar.nic.in [1]

Bhagni Pratham Khanda is a census town in Koch Bihar district in the state of West Bengal, India.

Geography

Bhagni Pratham Khanda is located at 26°07′N 89°28′E.[2]

Demographics

As of 2001 India census,[3] Bhangri Pratham Khanda had a population of 4113. Males constitute 51% of the population and females 49%. Bhangri Pratham Khanda has an average literacy rate of 69%, higher than the national average of 59.5%; with male literacy of 75% and female literacy of 63%. 12% of the population is under 6 years of age.

References

[1] http://coochbehar.nic.in

[2] "Yahoo maps location of Bhangri Pratham Khanda" (http://maps.yahoo.com/#mvt=m&lat=26.112591&lon=89.469994&zoom=15& q1=Bhangri%20Pratham%20Khanda). Yahoo maps. . Retrieved 2008-12-25.

[3] "Census of India 2001: Data from the 2001 Census, including cities, villages and towns (Provisional)" (http://web.archive.org/web/ 20040616075334/http://www.censusindia.net/results/town.php?stad=A&state5=999). Census Commission of India. Archived from the original (http://www.censusindia.net/results/town.php?stad=A&state5=999) on 2004-06-16. . Retrieved 2008-11-01.

Scheduled_castes_and_scheduled_tribes

The **Scheduled Castes** (SCs), also known as the **Dalit**, and the **Scheduled Tribes** (STs) are two groupings of historically disadvantaged people that are given express recognition in the Constitution of India. During the period of British rule in the Indian sub-continent they were known as the **Depressed Classes**.

The Scheduled Castes and Scheduled Tribes make up around 15% and 7.5% respectively of the population of India, or around 24% altogether, according to the 2001 Census.[1] The proportion of Scheduled Castes and Scheduled Tribes in the country's population has steadily risen since independence in 1947.

The *Constitution (Scheduled Castes) Order, 1950* lists 1,108 castes across 25 states in its First Schedule,[2] while the *Constitution (Scheduled Tribes) Order, 1950* lists 744 tribes across 22 states in its First Schedule.[3]

Since Independence, the Scheduled Castes have benefited by the "Reservation" policy. This policy became an integral part of the Constitution through the efforts of Dr. Bhimrao Ambedkar, regarded as the father of the Indian constitution, who participated in Round Table Conferences and fought for the rights of the Depressed Classes. The Constitution lays down general principles for the policy of affirmative action for the SCs and STs.

History

From the 1850s these communities were loosely referred to as the "Depressed Classes". The early part of the 20th century saw a flurry of activity in the British Raj to assess the feasibility of responsible self-government for India. The Morley-Minto Reforms Report, Montagu–Chelmsford Reforms Report, and the Simon Commission were some of the initiatives that happened in this context. One of the hotly contested issues in the proposed reforms was the reservation of seats for the "depressed" classes in provincial and central legislatures.

In 1935 the British passed the Government of India Act 1935, designed to give Indian provinces greater self-rule and set up a national federal structure. Reservation of seats for the Depressed Classes was incorporated into the act, which came into force in 1937. The Act brought the term "Scheduled Castes" into use, and defined the group as including "such castes, races or tribes or parts of groups within castes, races or tribes, which appear to His Majesty in Council to correspond to the classes of persons formerly known as the 'Depressed Classes', as His Majesty in Council may prefer". This discretionary definition was clarified in *The Government of India (Scheduled Castes) Order, 1936,* which contained a list, or Schedule, of castes throughout the British administered provinces.

After independence, the Constituent Assembly continued the prevailing definition of Scheduled Castes and Tribes, and gave (via articles 341, 342) the President of India and Governors of states responsibility to compile a full listing of castes and tribes, and also the power to edit it later as required. The actual complete listing of castes and tribes was made via two orders *The Constitution (Scheduled Castes) Order, 1950,*[4] and *The Constitution (Scheduled Tribes) Order, 1950*[5] respectively.

Constitutional framework for safeguarding of interests

The Constitution provides a framework with a three pronged strategy [6] to improve the situation of SCs and STs.

1. Protective Arrangements - Such measures as are required to enforce equality, to provide punitive measures for transgressions, to eliminate established practices that perpetuate inequities, etc. A number of laws were enacted to operationalize the provisions in the Constitution. Examples of such laws include The Untouchability Practices Act, 1955, Scheduled Caste and Scheduled Tribe (Prevention of Atrocities) Act, 1989, The Employment of Manual scavengers and Construction of Dry Latrines (Prohibition) Act, 1993, etc.
2. Affirmative action - provide positive preferential treatment in allotment of jobs and access to higher education, as a means to accelerate the integration of the SCs and STs with mainstream society. Affirmative action is also popularly referred to as Reservation.
3. Development - Provide for resources and benefits to bridge the wide gap in social and economic condition between the SCs/STs and other communities.

National commissions

To effectively implement the various safeguards built into the Constitution and other legislations, the Constitution, under Articles 338 and 338A, provides for two statutory commissions - the **National Commission for Scheduled Castes**, and **National Commission for Scheduled Tribes**.

History

In the original Constitution, Article 338 provided for a Special Officer, called the Commissioner for SCs and STs, to have the responsibility of monitoring the effective implementation of various safeguards for SCs/STs in the Constitution as well as other related legislations and to report to the President. To enable efficient discharge of duties, 17 regional offices of the Commissioner were set up all over the country.

In the meanwhile there was persistent representation for a replacement of the Commissioner with a multi-member committee. It was proposed that the 48th Amendment to the Constitution be made to alter Article 338 to enable said proposal. While the amendment was being debated, the Ministry of Welfare issued an administrative decision to establish the Commission for SCs/STs as a multi-member committee to discharge the same functions as that of the Commissioner of SCs/STs. The first commission came into being in August 1978. The functions of the commission were modified in September 1987 to advise Government on broad policy issues and levels of development of SCs/STs.

In 1990 that the Article 338 was amended to give birth to the statutory National Commission for SCs and STs via the *Constitution (Sixty fifth Amendment) Bill, 1990.*[7] The first Commission under the 65th Amendment was constituted in March 1992 replacing the Commissioner for Scheduled Castes and Scheduled Tribes and the Commission set up under the Ministry of Welfare's Resolution of 1989.

In 2002, the Constitution was again amended to split the National Commission for Scheduled Castes and Scheduled Tribes into two separate commissions - the National Commission for Scheduled Castes and the National Commission for Scheduled Tribes.

Distribution

According to the 61st Round Survey of the NSSO, almost nine-tenths of Buddhists in India belonged to scheduled castes of the Constitution while one-third of Christians belonged to scheduled tribes. Major part of scheduled castes were Hindus by religion but belonged to castes and tribes having low population. The Sachar Committee report of 2006 also confirmed that members of scheduled castes and tribes of India are not exclusively adherents of Hinduism.

Religion	Scheduled Caste	Scheduled Tribe
Buddhism	89.50%	7.40%
Christianity	9.00%	32.80%
Sikhism	17.0%	0.90%
Hinduism	22.20%	9.10%
Gond	-	15.90%
Jainism	-	2.60%
Islam	0.80%	0.50%

Scheduled Caste Sub-Plan

The strategy of Scheduled Castes Sub-Plan (SCSP) of 1979 is an important intervention that mandates a planning process for social, economic, and educational development of Scheduled Castes and for improvement in their working and living conditions. It is an umbrella strategy that ensures flow of targeted financial and physical benefits from general sectors of development for the benefit of Scheduled Castes. Under this strategy, population.[8] It entails targeted flow of funds and associated benefits from the annual plan of States / Union Territories (UTs) at least in proportion to the SC population i.e. 16 % in the total population of the country / the particular state. Presently, 27 States / UTs having sizeable SC populations are implementing Scheduled Castes Sub-Plan. Although the Scheduled Castes population, according to 2001 Census, was 16.66 crores constituting 16.23% of the total population of India, the allocations made through SCSP in recent years have been much lower than the population proportion. Table hereafter provides the details of total State Plan Outlay, flow to Scheduled Castes Sub-Plan (SCSP) as reported by the State / UT Governments for the last few years especially since the present UPA government is in power at the

2004–2005	108788.9	17656	2065.38	11.06	68.3	5591
2005–2006	136234.5	22111	16422.63	12.05	74.3	5688
2006–2007	152088	24684	21461.12	14.11	86.9	3223
2007-2008*	155013.2	25159	22939.99	14.80	91.2	2219

- Information in respect of 14 States/UTs only and as on 31-12- 2007

Source: Network for Social Accountability (NSA) http://nsa.org.in

Prominent Personalities of SC/STs Community

- Guru Ravidas,North Indian Sant mystic of the bhakti movement
- Khusro Khan, or Khusru or Khusraw Khan was a medieval Indian military leader, and ruler of Delhi, as Sultan Nasir-ud-din, for a short period of time.He was a Dalit (Parwari-Mahar) caste from Gujrat. He converted to Islam from Hinduism at the time of his capture.[9] He was a untouchable in his own religion, but became a first Hindu to sit on the throne of Delhi.
- K. R. Narayanan, tenth President of India

- * Babu Jagjivan Ram, former Deputy Prime Minister of India
- B. R. Ambedkar, jurist, political leader, writer, father of Indian Constitution
- K. G. Balakrishnan, former Chief Justice of India
- Sushilkumar Shinde, Current Cabinet Minister for Power
- Prof. Nibaran Chandra Laskar, MP, Indian Parliament, was a dalit leader in Bengal and Assam.
- Mayavati, Chief Minister of Uttar Pradesh.
- Birsa Munda, Indian independence advocate, tribal leader and folk hero
- Damodaram Sanjivayya (1921–1972) (First dalit Chief Minister of a state in India and first dalit President of Indian National Congress party)
- Kanshi Ram, founder of Bahujan Samaj Party
- Dr. Mahendra Chandra Patni, a dalit leader and prominent scientist who has got the gold medal in LMP (the then MBBS) in 1923-24 batch from Berry-White School of Medicine, Dibrugarh, Assam, British India.
- D.Raja, Member of Rajyasabha,National Secratory for Communist Party of India
- G. M. C. Balayogi, dalit speaker, Lok Sabha,
- Ajit Jogi, first chief minister of the state of Chhattisgarh, India
- Shibu Soren, Ex Chief Minister of Jharkhand state in India
- Meira Kumar, Indian politician and Member of Parliament, Speaker of Lok Sabha
- S. Ashok Kumar, Judge Madras High Court and High Court of Andhra Pradesh
- Ram Vilas Paswan, the president of the Lok Janshakti Party, political party
- Bangaru Laxman, former President of Bharatiya Janata Party (BJP)
- Lala Ram Ken, Member of Parliament (7th and 8th), India
- Vinod Kambli, Indian cricketer
- Vinoo Mankad, Indian cricketer, He played in 44 Tests for India
- Thol. Thirumavalavan, Member of Parliament, The founder president of Viduthalai Chiruthaikal Katchi, Tamil Nadu
- Ilaiyaraaja, a noted music director and composer, Ilaiyaraaja is also a instrumentalist, conductor, and a songwriter
- E. Ponnuswamy, former M.O.S. Petroleum India.
- M.E.Loganathan, Municipal Commissioner, Government of Tamil Nadu.
- Damodar Raja Narasimha - Deputy Chief Minister of Andhra Pradesh
- Dr. J. Geeta Reddy - Leader of the House in the Legislative Assembly AP
- Amarjeet Bhagat MLA Sitapur
- Faguni Ram, Member of Parliament and Minister of State
- K. S. R. Murthy IAS, Former MP, Lok Sabha
- Prem Singh - MLA
- Late Lahori Ram Economic Development Commissioner California State and Founder member Guru Ravidass Sikh Gurdwara, Pittusburg
- Ram Lakha Former Mayor of Coventry
- Sardar Lakhbir Singh First Sikh Mayor Of Luton
- Giani Ditt Singh Ji Founder of Singh Sabha Movement
- Dr. Baldev Singh Sher First Dalit (Ravidasia/Ramdasia Sikh) Medical Graduate from Glasgow in 1910 and son of Giani Ditt Singh Ji
- Shaeed Baba Sangat Singh Ji Martyr in the Battle of Chamkaur Sahib
- Johnny Lever (Janumala John Prakasa Rao) - Famous Bollywood comedian, born in Vusullapalli near Kanigiri, Prakasm dt, AP.
- Betha Sudhakar ("Pichha kottudu sudhakar") - Popular Comedian & Character Artist in Tollywood
- Lankapalli Bullayya(1918–1992), former VC Andhra University(1968–74); first dalit to become the Vice-Chancellor of a university in India

- Late Shri Ram Ratan Ram— Member of Parliament (1984–1989)
- Dr.M.Velusamy (1973) is well known Social Science Scholar from Tamil Nadu. First Dalit Scholar Who has awarded his PhD at Madras Institute of Development Studies (MIDS), Chennai. His thesis entitled on Indian Constitution and Dalit Welfare : A Study of Tamil Nadu, 1950-2005. Published books on topic related to Dalits Periyar Dravidian Politics in Tamil Nadu.
- Jwala Prasad Kureel- MP of 6th Lok Sabha, Affiliated to Janata Party serving Ghatampur (UP) Lok Sabha Constituency
- Arun Anand- Former Scholar BTech Mech, MBA IIT Delhi, Social Worker, Running NGO for Dalit and poor in bangalore

See also

- List of Scheduled Tribes in India
- Scheduled Caste and Scheduled Tribe (Prevention of Atrocities) Act, 1989
- List of Sudra Hindu Saints
- Dalit saints of Hinduism
- Forward caste
- Other Backward Class
- Dalit Christian

Notes

[1] Census of India - India at a Glance : Scheduled Castes & Scheduled Tribes Population (http://www.censusindia.gov.in/ Census_Data_2001/India_at_Glance/scst.aspx)
[2] Text of the *Constitution (Scheduled Castes) Order, 1950*, as amended (http://lawmin.nic.in/ld/subord/rule3a.htm)
[3] Text of the *Constitution (Scheduled Tribes) Order, 1950*, as amended (http://lawmin.nic.in/ld/subord/rule9a.htm)
[4] THE CONSTITUTION (SCHEDULED CASTES) ORDER, 1950]1 (http://lawmin.nic.in/ld/subord/rule3a.htm)
[5] 1THE CONSTITUTION (SCHEDULED TRIBES) (http://lawmin.nic.in/ld/subord/rule9a.htm)
[6] http://nhrc.nic.in/Publications/reportKBSaxena.pdf
[7] "Constitution of India as of 29 July 2008" (http://lawmin.nic.in/coi/coiason29july08.pdf). *The Constitution Of India*. Ministry of Law & Justice. . Retrieved 13 April 2011.
[8] http://www.planningcommission.nic.in/plans/stateplan/scp&tsp/noteguidelinesFor.doc

External links

- Jobs for Tribals (http://tribaljobsindia.blogspot.com/)
- Dalit Indian Chamber of Commerce & Industry (http://www.dicci.org/index.html)
- Dalit and Adivasi Student Portal (http://www.scststudents.org)
- Rise of Dalit businessmen (http://ibnlive.in.com/news/dalit-becomes-entrepreneur-without-use-of-quotas/ 189929-3.html)
- Organization for SC & ST Govt Employees (http://www.ajjaks.com/)

Delimitation_Commission_of_India

Delimitation commission or **Boundary commission** of India is a Commission established by Government of India under the provisions of the Delimitation Commission Act. The main task of the commission is to redraw the boundaries of the various assembly and Lok Sabha constituencies based on a recent census. The representation from each state is not changed during this exercise. However, the number of SC and ST seats in a state are changed in accordance with the census.

The Commission is a powerful body whose orders cannot be challenged in a court of law. The orders are laid before the Lok Sabha and the respective State Legislative Assemblies. However, modifications are not permitted.

Past Commissions

Delimitation commissions have been set up four times in the past - In 1952, 1963, 1973 and 2002 under Delimitation Commission acts of 1952, 1962, 1972 and 2002.

The government had suspended delimitation in 1976 until after the 2001 census so that states' family planning programmes would not affect their political representation in the Lok Sabha. This had led to wide discrepancies in the size of constituencies, with the largest having over three million electors, and the smallest less than 50,000. [1]

Commission of 2002

The recent delimitation commission was set up on 12 July 2002 after the 2001 census with Justice Kuldip Singh, a retired Judge of the Supreme Court of India as its Chairperson. The Commission has submitted its recommendations. On December 2007, the Supreme Court of India on a petition issued notice to the central government for non implementation. On 4 January 2008, the CCPA decided to implement the order from the Delimitation commission [2] . The recommendations of the delimitation commission was approved by the President, Pratibha Patil on 19 February 2008. This means that all future elections in India for states covered by the commission will be held under the newly formed consistencies[3] .

The assembly elections in Karnataka which were conducted in three phases in May 2008 is the first one to use the new boundaries as drawn by the 2002 delimitation commission.[4]

References

[1] Election Commission of India (http://www.eci.gov.in/ElectoralSystem/the_function.asp#howconstituency)

[2] Delimitation process now gets CCPA nod- Politics/Nation-News-The Economic Times (http://economictimes.indiatimes.com/News/ PoliticsNation/Delimitation_process_now_gets_CCPA_nod/articleshow/2673204.cms)

[3] Delimitation notification comes into effect - The Hindu 20 February 2008 (http://www.hindu.com/2008/02/20/stories/ 2008022058631200.htm)

[4] Delimitation may kick off with Karnataka (http://www.financialexpress.com/news/Delimitation-may-kick-off-with-Karnataka/257699/)

External links

- Interesting facts about Lok Sabha Constituencies (http://www.rediff.com/election/2004/apr/12espec.htm)
- Karnataka assembly polls (http://search.eci.gov.in/ae_2008e/)

Dinhata_I_(community_development_block)

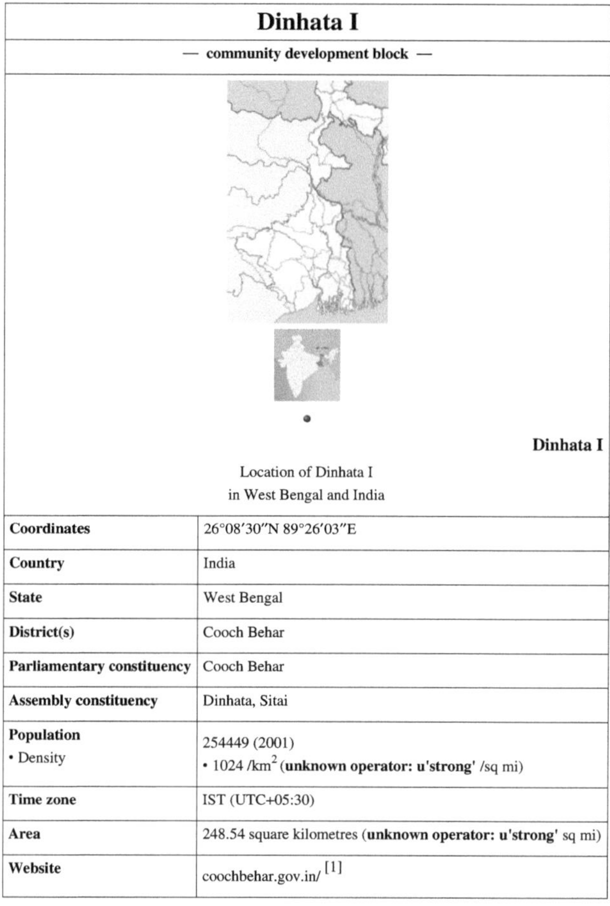

Dinhata I	
— community development block —	
Coordinates	26°08′30″N 89°26′03″E
Country	India
State	West Bengal
District(s)	Cooch Behar
Parliamentary constituency	Cooch Behar
Assembly constituency	Dinhata, Sitai
Population • Density	254449 (2001) • 1024 /km^2 (**unknown operator: u'strong'** /sq mi)
Time zone	IST (UTC+05:30)
Area	248.54 square kilometres (**unknown operator: u'strong'** sq mi)
Website	coochbehar.gov.in/ [1]

Location of Dinhata I
in West Bengal and India

Dinhata I (community development block) (Bengali: দিনহাটা I সমষ্টি উন্নয়ন ব্লক) is an administrative division in Dinhata subdivision of Cooch Behar district in the Indian state of West Bengal. Dinhata police station serves this block. Headquarters of this block is at Dinhata. There is one census town in this block: Bhangri Pratham Khanda.[2][3]

Geography

Petla, one of the constituenct panchayats of the block, is located at 26°08′30″N 89°26′03″E.

Dinhata I community development block has an area of 248.54 km^2.[3]

Gram panchayats

Gram panchayats of Dinhata I block/ panchayat samiti are: Bara Atiabari I, Bara Atiabari II, Bara Soulmari, Bhetaguri I, Bhetaguri II, Dinhata Village I, Dinhata Village II, Gitaldaha I, Gitaldaha II, Gosainmari I, Gasainmari II, Matalhat, Okrabari, Petla, Putimari I and Putimari II.[4]

Demographics

As per 2001 census, Dinhata I block had a total population of 254,449, out of which 130,656 were males and 123,793 were females. Dinhata I block registered a population growth of 6.04 per cent during the 1991-2001 decade. Decadal growth for the district was 14.15 per cent.[3] Decadal growth in West Bengal was 17.84 per cent.[5]

Literacy

Dinhata I block had a literacy rate of 64.16 per cent - males 73.27 per cent and females 54.55 per cent.[4]

References

[1] http://coochbehar.gov.in/

[2] "Contact details of Block Development Officers" (http://wbprd.gov.in/html/asp/bdo_contact.asp?cd=DL). *Cooch Behar district*. West Bengal Government. . Retrieved 2011-03-20.

[3] "Census of India 2001, Provisional Population Totals, West Bengal, Table - 4" (http://web.archive.org/web/20070927041813/http://www.wbcensus.gov.in/DataTables/02/Table4_3.htm). *Cooch Behar District (03)*. Government of West Bengal. . Retrieved 2012-04-12.

[4] "Relation between Blocks & Gram Panchayats (GPs)" (http://coochbehar.nic.in/htmfiles/District_Administration.html). Cooch Behar District Administration. . Retrieved 2011-03-20.

[5] "Provisional Population Totals, West Bengal. Table 4" (http://web.archive.org/web/20070927041813/http://www.wbcensus.gov.in/DataTables/02/Table4_1.htm). *Census of India 2001*. Census Commission of India. . Retrieved 2012-04-12.

Boundary_delimitation

Boundary delimitation, or simply **delimitation**, is the term used to describe the drawing of boundaries, but is most often used to describe the drawing of electoral boundaries, specifically those of precincts, states, counties or other municipalities.[1] Often this takes places in democracies; in this context it can be called redistribution in order to prevent unbalance of population across districts.[1] Unbalanced or discriminatory delimitation is called "gerrymandering."[2] Though there are no internationally agreed processes which guarantee fair delimitation, several organizations, such as the Commonwealth Secretariat, the European Union and the International Foundation for Electoral Systems have proposed guidelines for effective delimitation.

In international law, boundary delimitation is sometimes referred to as **national delimitation**. Most specifically this refers to a process of legally establishing the outer limits ("borders") of a state within which full territorial or functional sovereignty is exercised.[3] Occasionally this is used when referring to the maritime boundaries as well, in this case called **maritime delimitation**.

Democratic delimitation

Methods

Countries delimit electoral districts in different ways.[1] Sometimes these are drawn based on traditional boundaries, sometimes based on the physical characteristics of the region and, often, the lines are drawn based on the social, political and cultural contexts of the area.[1] This may need to be done in any form of electoral system even though it is primarily done for plurality or majority electoral system.[1]

These processes of boundary delimitation can have a variety of legal justifications. Often, because of the powerful effects this process can have on constituencies, the legal framework for delimitation is specified in the constitution of a country.[4] The Institute for Democracy and Electoral Assistance (IDEA) recommends the following pieces of information be included in this legal framework:[4]

- The frequency of such determination;
- The criteria for such determination;
- The degree of public participation in the process;
- The respective roles of the legislature, judiciary and executive in the process;and
- The ultimate authority for the final determination of the electoral units.

Established democracies

Delimitation is used in the United States and Commonwealth countries. This is called redistricting or redistribution respectively. In these countries non-partisan commissions draw new districts based on the distribution of population according to a census.

International Standards

A number of international organizations including the Organization for Security and Co-operation in Europe, the European Commission for Democracy Through Law (the Venice Commissio), the Commonwealth Secretariat, and the Electoral Institute of Southern Africa (EISA) have established standards which their members are encouraged to prescribe to.[2] Among these standards the International Foundation for Electoral Systems (IFES) lists the most common as being Impartiality, Equality, Representativeness, Non-Discrimination and Transparency.[2]

Venice Commission

As part of its report, *European Commission for Democracy Through Law: Code of Good Practice in Electoral Matters, Guidelines and Explanatory Reports adopted October 2002*, the Venice Commission proposed the following guidelines:[2]

2.2 Equal voting power: seats must be evenly distributed between the constituencies.

i. This must at least apply to elections to lower houses of parliament and regional and local elections:

ii. It entails a clear and balanced distribution of seats among constituencies on the basis of one of the following allocation criteria: population, number of resident nationals (including minors), number of registered voters, and possibly the number of people actually voting. An appropriate combination of these criteria may be envisaged.

iii. The geographical criterion and administrative, or possibly even historical,boundaries may be taken into consideration.

iv. The permissible departure from the norm should not be more than 10%, and should certainly not exceed 15% except in special circumstances (protection of a concentrated minority, sparsely populated administrative entity).

v. In order to guarantee equal voting power, the distribution of seats must be reviewed at least every ten years, preferably outside election periods.

vi. With multimember constituencies, seats should preferably be redistributed without redefining constituency boundaries, which should, where possible, coincide with administrative boundaries.

vii. When constituency boundaries are redefined—which they must be in a single-member system—it must be done:

- impartially;

- without detriment to national minorities;

- taking account of the opinion of a committee, the majority of whose members are independent; this committee should preferably include a geographer, a sociologist, and a balanced representation of the parties and, if necessary, representatives of national minorities.

Commonwealth Secretariat

In the publication *Good Commonwealth Electoral Practices: A Working Document, June 1997,* the Commonwealth Secretariat identifies the following practices as necessary for proper delimitation:[2]

20. The delimitation of constituency boundaries is a function occasionally performed by an election commission or otherwise by an independent boundaries commission, and in some cases after a population census.

21. General principles guiding the drawing of constituency boundaries include community of interest, convenience, natural boundaries, existing administrative boundaries and population distribution, including minority groups. There should be no scope for any "gerrymandering", and each vote should, to the extent possible, be afforded equal value or weight, in recognition of the democratic principle that all those of voting age participate equally in the ballot.

22. It is important that the general public play a part in the whole process and that the political parties also have an opportunity to respond to proposals before they are finalized. Where the size of a particular constituency is markedly out of line with the target "quota" of voters per seat, the reasons should be capable of being readily understood by both the parties and the general public.

IFES

In her study sponsored by the International Foundation for Electoral Systems, Dr. Lisa Handley recommends the following considerations:[2]

1. population density
2. ease of transportation and communication
3. geographic features
4. existing patterns of human settlement
5. financial viability and administrative capacity of electoral area
6. financial and administrative consequences of boundary determination
7. existing boundaries
8. community of interest

Also, she suggests that the process should:[2]

- be managed by an independent and impartial body that is representative of society, comprising persons with the appropriate skills;
- be conducted on the basis of clearly identified criteria such as population, distribution, community of interest, convenience, geographical features and other natural or administrative boundaries;
- be made accessible to the public through a consultation process;
- be devoid of manipulation of electoral boundaries to favour political groups or political interests;
- be conducted by one body;
- include all spheres of government, both national and local.

National delimitation

The negotiations surrounding the modification of a states borders is called National delimitation. This event often takes place as part of the negotiations seeking to end a conflict over resource control, popular loyalties, or political interests.

Maritime delimitation

The term Maritime delimitation is a form of national delimitation that can be applied to the disputes between nations over maritime claims. An example is found at Maritime Boundary Delimitation in the Gulf of Tonkin [5]. In international politics, the Division for Ocean Affairs and the Law of the Sea, Office of Legal Affairs, United Nations Secretariat is responsible for the collection of all claims to territorial waters.[6]

See also

For further examples of legislative delimitation:

- Delimitation Commission of India

For further elaboration of this concept of national delimitaiton see

- National delimitation in the Soviet Union on the creation of territorial units based on ethnicity in the USSR;
- Nation-building on the processes of creating or strengthening national identity within national territorial limits.
- Sugauli Treaty

For further examples of the concept maritime delimitation:

- Maritime delimitation between Romania and Ukraine
- Georges Bank
- List of maritime boundary treaties

References

[1] Overview of Boundary Delimitation (http://aceproject.org/ace-en/topics/bd/bd10) ACE: The Electoral Knowledge Center. Accessed July 09, 2008.

[2] Challenging the Norms and Standards of Election Administration Boundary Delimitation (http://www.ifes.org/publication/ 505d087c7a033c8563e67b9fdd45cd78/4 IFES Challenging Election Norms and Standards WP BNDEL.pdf). IFES, 2007. Accessed July 09, 2009.

[3] G.J. Tanja, Ministry of Foreign Affairs, The Netherlands, comment in I.F. Dekker, H.H.G. Post, and T.M.C. Asser, *The Gulf War of 1980-1988: The Iran-Iraq War in International Legal Perspective*, Martinus Nijhoff Publishers (1992), pp.44-45.

[4] Boundary delimitation, districting ordefining boundaries of electoral units (http://www.idea.int/publications/ies/upload/4.Boundary delimitation, districting or defining boundaries of electoral units.pdf) a chapter from *International Electoral Standards: Guidelines for reviewing the legal framework of elections*. Institute for Democracy and Electoral Assistance. Accessed July 21, 2009

[5] http://www.southchinasea.org/docs/Keyuan,%20Zou,%20Tonkin%20Gulf%20Maritime%20Boundary%20Delimitation.pdf

[6] "Maritime Space: Maritime Zones and Maritime Delimitation" (http://www.un.org/Depts/los/LEGISLATIONANDTREATIES/). UN. . Retrieved Nov 15, 2009.

External links

- UN database of Maritime delimitation treaties (http://www.un.org/Depts/los/ LEGISLATIONANDTREATIES/)

Cooch_Behar_(Lok_Sabha_constituency)

Existence	1957-present
Reservation	Reserved for SC
Current MP	Nripendra Nath Roy
Party	Forward Bloc
Elected Year	2009
State	West Bengal
Total Electors	1,341,455
Assembly Constituencies	Mathabhanga (SC) Coochbehar Uttar Coochbehar Dakshin Sitalkuchi (SC) Sitai (SC) Dinhata Natabari

Cooch Behar (Lok Sabha constituency) (Bengali: কোচবিহার লোকসভা কেন্দ্র) is one of the 543 parliamentary constituencies in India. The constituency centres on Cooch Behar in West Bengal. All the seven assembly segments of No. 1 Cooch Behar (Lok Sabha constituency) are in Cooch Behar district. The seat is reserved for scheduled castes.

Assembly segments

As per order of the Delimitation Commission in respect of the delimitation of constituencies in the West Bengal, parliamentary constituency no. 1 Coochbehar, reserved for Scheduled castes (SC), is composed of the following segments from 2009:[1]

- Mathabhanga (SC) (assembly constituency no. 2)
- Coochbehar Uttar (assembly constituency no. 3)
- Coochbehar Dakshin (assembly constituency no. 4)
- Sitalkuchi (SC) (assembly constituency no. 5)
- Sitai (SC) (assembly constituency no. 6)
- Dinhata (assembly constituency no. 7)
- Natabari (assembly constituency no. 8)

The area under the Mathabhanga subdivision of the Cooch Behar district will constitute the assembly constituencies of Mathabhanga and Sitalkuchi, whereas the area under the Dinhata subdivision will form the constituencies of Dinahata and Sitai.[1] The area under Cooch Behar Sadar subdivision will form Cooch Behar Uttar, Cooch Behar Dakshin and Natabari constituencies, though Natabari will contain gram panchayats from Tufanganj subdivision also.[1]

In 2004 Cooch Behar Lok Sabha constituency was composed of the following assembly segments:[2]

- Sitalkuchi (SC) (assembly constituency no. 2)
- Mathabhanga (SC) (assembly constituency no. 3)
- Cooch Behar North (assembly constituency no. 4)
- Cooch Behar West (assembly constituency no. 5)
- Sitai (assembly constituency no. 6)

- Dinhata (assembly constituency no. 7)
- Natabari (assembly constituency no. 8)

Election results

General Election, 2009: Cooch Behar[3]				
Party	**Candidate**	**Votes**	**%**	**±%**
Forward Bloc	Nripendra Nath Roy	500,677	44.66	
Trinamool Congress	Arghya Roy Pradhan	466,928	41.65	
BJP	Bhabendra Nath Barman	65,325	5.83	
Independent	Bangshi Badan Barman	37,226	3.32	
BSP	Niranjan Barman	22,925	2.04	
Independent	Hitendra Das	11,374	1.01	
AMB	Dalendra Roy	6,486	0.58	
Independent	Nubash Barman	3,737	0.33	
RPI	Harekrishna Sarkar	3,405	0.30	
Independent	Krishna Kanta Barman	2,960	0.26	
	Turnout	1,121,043	84.35	
Forward Bloc **hold**		**Swing**		

Indian general election, 2009

West Bengal summary

Party	Seats won	Seat change
Trinamool Congress	19	▲ 18
Indian National Congress	6	▲ 0
Socialist Unity Centre of India (Communist)	1	▲ 1
Communist Party of India (Marxist)	9	▼ 17
Communist Party of India	2	▼ 1
Revolutionary Socialist Party	2	▼ 1
Forward bloc	2	▼ 1
Bharatiya Janata Party	1	▲ 1

Source: List of successful candidates in General Elections 2009 to the 15th Lok Sabha [4]

Statitical Report on General Elections 2004 to the 14th Lok Sabha [5]

Results prior to 2009

Results of elections held prior to 2009 are summarised below:

1951-1971

In 1951, Upendra Nath Barman, Birendra Nath Katham and Amiya Kanta Basu, all of Congress, won the North Bengal seat.[6] The winners from Cooch Behar for subsequent elections are shown below.

Year	Winner	Party
1957[7]	Santosh Banerjee	Congress
	Upendra Nath Barman	Congress
1958 (by-election)[8]	Nalini Ranjan Ghosh	Congress
1962[9]	Debendranath Karjee	Forward Bloc
1967[10]	Benoy Krishna Daschowdhury	Congress
1971[11]	Benoy Krishna Daschowdhury	Congress

1977-2004

Year	Winner		Runner-up	
	Candidate	Party	Candidate	Party
1977[12]	Amreandranath Roy Pradhan	Forward Bloc	Benoy Krishna Daschowdhury	Indian National Congress
1980	Amreandranath Roy Pradhan	Forward Bloc	Ambika Charan Roy	Indian National Congress (I)
1984	Amreandranath Roy Pradhan	Forward Bloc	Prasenjit Barman	Indian National Congress
1989	Amreandranath Roy Pradhan	Forward Bloc	Sabita Roy	Indian National Congress
1991	Amreandranath Roy Pradhan	Forward Bloc	Sabita Roy	Indian National Congress
1996	Amreandranath Roy Pradhan	Forward Bloc	Sabita Roy	Indian National Congress
1998	Amreandranath Roy Pradhan	Forward Bloc	Gobinda Roy	Forward Bloc (Socialist)
1999	Amreandranath Roy Pradhan	Forward Bloc	Ambika Charan Ray	All India Trinamool Congress
2004	Hiten Barman	All India Forward Bloc	Girindra Nath Barman	All India Trinamool Congress

See also

- List of Constituencies of the Lok Sabha

References

[1] "Delimitation Commission Order No. 18" (http://ceowestbengal.nic.in/news_pdf/gazette123.pdf). *Table B − Extent of Parliamentary Constituencies*. Government of West Bengal. . Retrieved 2009-05-27.

[2] "Statistical Report on General Elections, 2004 to the 14th Lok Sabha" (http://eci.nic.in/eci_main/statisticalreports/LS_2004/Vol_III_LS_2004.pdf). *Volume III Details For Assembly Segments Of Parliamentary Constituencies*. Election Commission of India. . Retrieved 2010-10-01.

[3] "Constituency Wise Detailed Results" (http://eci.gov.in/eci_main/archiveofge2009/Stats/VOLI/25_ConstituencyWiseDetailedResult.pdf). *Election Commission of India, General Elections, 2009 (15th LOK SABHA) Section - West Bengal*. Election Commission of India. . Retrieved 2011-01-15.

[4] http://eci.nic.in/eci_main/archiveofge2009/Stats/VOLI/11_ListOfSuccessfulCandidate.pdf

[5] http://eci.nic.in/eci_main/StatisticalReports/LS_2004/Vol_I_LS_2004.pdf

[6] "General Elections, India, 1951" (http://eci.nic.in/eci_main/StatisticalReports/LS_1951/VOL_1_51_LS.PDF). Election Commission. .
Retrieved 2012-03-13.

[7] "General Elections, India, 1957" (http://eci.nic.in/eci_main/StatisticalReports/LS_1957/Vol_I_57_LS.pdf). Election Commission. .
Retrieved 2012-03-13.

[8] "Bengal Congress MP, MLA" (http://www.wbpcc.org/congmpmla.htm). West Bengal Pradesh Congress Committee. .

[9] "General Elections, India, 1962" (http://eci.nic.in/eci_main/StatisticalReports/LS_1962/Vol_I_LS_62.pdf). Election Commission. .
Retrieved 2012-03-13.

[10] "General Elections, India, 1967" (http://eci.nic.in/eci_main/StatisticalReports/LS_1967/Vol_I_LS_67.pdf). Election Commission. .
Retrieved 2012-03-13.

[11] "General Elections, India, 1971" (http://eci.nic.in/eci_main/StatisticalReports/LS_1971/Vol_I_LS71.pdf). Election Commission. .
Retrieved 2012-03-13.

[12] "1 – Cooch Behar Parliamentary Constituency" (http://www.eci.nic.in/eci_main/electionanalysis/GE/PartyCompWinner/S25/
partycomp01.htm). *Partywise Comparison since 1977*. Election Commission of India. . Retrieved 2010-10-01.

Article Sources and Contributors

Dinhata_subdivision *Source*: http://en.wikipedia.org/w/index.php?title=Dinhata_subdivision *Contributors*: Chandan Guha, GDibyendu, Sadads, 4 anonymous edits

Cooch_Behar_district *Source*: http://en.wikipedia.org/w/index.php?title=Cooch_Behar_district *Contributors*: Ahoerstemeier, Amartyabag, Antorjal, Antur, Apurbasen, Aryasanyal, AvicAWB, Awiseman, Aymatth2, Bagchivivek, Bhadani, Bhadra105, Biplab.K.M, Biruitorul, Bobblewik, Cdc, Chandan Guha, CrazyLegsKC, Crusoe8181, Cwilli201, Debasishkoley, Ebben, Ecnemalas, Ekabhishek, Faradayplank, GDibyendu, Gaius Cornelius, Gene Nygaard, Generalboss3, Grafen, Hetam, Hydkat, ImpuMozhi, Iridescent, Jeroen, JingaJenga, Jj137, Joy1963, Jpeeling, JustAGal, Kelisi, Kguneet, Kintetsubuffalo, Looper5920, Lucio Di Madaura, MK2, Martinp23, Master Of Ninja, Matt.T, Mikeblas, Mitsukai, Moonraker12, NastalgicCam, Nichalp, Otterlover, P.K.Niyogi, Partha1022, Pearle, PhnomPencil, Ragib, Raju14022012, Rixon45, Rjwilmsi, Sadads, Seaoneil, Size J Battery, SpacemanSpiff, Spellmaster, Tabletop, Tony164, Trakesht, Vanjagenije, Vrenator, Waggers, Xufanc, आशिष भटनागर, 99 anonymous edits

West_Bengal *Source*: http://en.wikipedia.org/w/index.php?title=West_Bengal *Contributors*: *drew, 5 albert square, Aaroncrick, Abdussamad77, Abhisek2091, Abhowmick, Acsenray, Admrboltz, AgnosticPreachersKid, Ahoerstemeier, Airunp, AjaxSmack, Ak2431989, Akkida, Al Silonov, AlexanderKaras, AlimanRuna, Alokchakrabarti, Aloke Kumar, Amartyabag, Ambuj.Saxena, Amitabdev, Amitsanyal, Amplitude101, Anchitk, Anclation, Andrewlp1991, Andyindia, Aneesharry, Aniketmg, Anirvan, Ankitbhatt, Annalise, Anshuman.jrt, Antorjal, Anwar saadat, Aotearoa, AralSea, Arnab kl4, Arnabik, Arnabik9, Arparag, Art LaPella, Aryasanyal, Ashowmega, Astrokid, Aurorion, Avenue X at Cicero, AvicAWB, Avinashv11, Avr0716ap, AxelBoldt, BD2412, Babuonwiki, Bapan25, Bbbl67, Begoon, Bharatveer, Biblbroks, Bijoy Krishna Das, Bility, Bl4ze013, Bless sins, Blue-Haired Lawyer, Bobblewik, Bonadea, BostonMA, Brewcrewer, Brighterorange, Buaidh, CJLL Wright, CLW, Cab.jones, Cactus.man, Caeruleancentaur, Caltas, CarTick, Cecil, Chandan Guha, Charles Matthews, Charlesdrakew, Chirag, Chrism, Cloudbound, Clt13, Combustion X, Conversion script, D, D6, DARTH SIDIOUS 2, DMS, DaGizza, Dark Shikari, Dasoumendu, Dav subrajathan.357, David matthews, Dawn Bard, Debasishkoley, Debjato, Deepak D'Souza, Deepika11, Deeptrivia, Dekimasu, Der Golem, Dewan357, Dipankan001, Discospinster, Dlohcierekim, Dn9ahx, Domino theory, Drmies, Drpickem, Drumguy8800, Dwaipayanc, East Bahamas, Ed Poor, El C, Elockid, Eluchil404, Enti342, Enviroboy, EoGuy, Epipelagic, Eras-mus, Eukesh, Expresswaytoparadise, Filemon, Fnielsen, Fnatrep, Fundamental metric tensor, GDibyendu, GSMR, Gaius Cornelius, Galoubet, Ganeshk, Gangulybiswarup, Gaurav38, Gene Nygaard, Generalboss3, Gimmetrow, Gman124, Goatse Rulz, Gogo Dodo, GoingBatty, Golbez, Gomada, Good Olfactory, Gopalagarwal11, Gppande, Graham87, Grenavitar, Grey Scaler, Ground Zero, Gurubrahma, Gzornenplatz, Hagedis, Haham hanuka, HeBhagawan, Hello2abir, Help Always, Hemanshu, Hindubengali, Hippietrail, Holy Ganga, Hometech, Hqb, Ian Pitchford, Improv, InMooseWeTrust, India Gate, Indon, Indranee, Interchange88, JForget, JaGa, Jac16888, Jagged 85, Jahangard, Japanese Searobin, Jauhienij, Jayantanth, Jeroje, Jethwarp, Jiminy900, Jimp, Jmgarg1, Jojit fb, Jon4562002, JonnyNYC90, Jonoikobangali, Jorunn, Jovianeye, Joy9, Joyson Prabhu, Justarijit, Kamalakardandu, Kannanwrites, Kartik2008, Kaushikhm, Kayau, Kbdank71, Kensplanet, Khazar, Kintetsubuffalo, Kiril Simeonovski, Kirti 1102, Kkm010, KnowledgeHegemony, Knutux, Koavf, Kozuch, Kristian Vangen, Kuaichik, Kungming2, Kurykh, KuwarOnline, Kwamikagami, Lexington1, Lightmouse, Ligulem, Livajo, Lkinkade, Logan, Logicalthinker33, Lollywood, Lord of the Goatse, LordGulliverofGalben, LordSimonofShropshire, LordSuryaofShropshire, MJCdetroit, Maddyr, Mainak, Mani1, Manoj nav, Martial75, Master Of Ninja, Materialscientist, Mattbr, Maurice45, Mayukh iitbombay 2008, Mboverload, Mddeath, MeekSaffron, MegaSloth, Mhchintoo, Michael Devore, Michaelmas1957, Mightymights, Mike R, Mike Rosoft, MikeLynch, Mild Bill Hiccup, Millahnna, Minusia, Mister Goat, Mkeane1019, Mkweise, Morwen, Mr Tan, Mr0t1633, Munci, Munita Prasad, Murphy dean, Murtasa, Mvp15, My76Strat, Nafsadh, Nakon, Naru12333, Nat Krause, Nataliehorsey, Nataraja, NawlinWiki, Neddyseagoon, NeelAbodh, Nichalp, Nigholith, Nikkul, Niteowlneils, Niteshpradhans, Northumbrian, Npeters22, Nunamiut, Obradovic Goran, OhanaUnited, Olivier, OneGuy, Oolong, Overvital1288, Oystertoadfish, P.K.Niyogi, Paleolithic1288, Patel24, Patrick, Paul-L, Pavel Vozenilek, Pax:Vobiscum, Pdutta81, Peoplespower, Petiatil, Pgk, Philyiverson3, PhnomPencil, PigFlu Oink, Piledhigheranddeeper, Pill, Pit, Planemad, Pollodiablowiki, Ppntori, Pradip ch, Pradiptaray, Priyatu, Prodego, Proofreader77, Proud Ho, Psubhashish, Pupunwiki, QuartierLatin1968, Quibik, Qwerta369, Qxz, R'n'B, RJFJR, Rabbabodrool, Radar signal, Ragib, Rahulghose, Raj Krishnamurthy, Rama's Arrow, Ramayan, Raul654, Rbha7, Recognizance, Redtigerxyz, Rehan.haider, RexNL, Riana, Rich Farmbrough, Rick Block, Ritwikbmca, Rjwilmsi, Robertgreer, Ronline, Rudroanik, Rueben lys, Saga City, Saimdusan, Sam Hocevar, SameerKhan, Samir, Sander123, SandyGeorgia, Saptarshivet, Saravask, Sardanaphalus, Sarvagnya, Saurabh.vinian, Savitr, Sayanmitra, Scjessey, Scope creep, Scottinglis, ScribeOO, Scriber, Sd31415, Sen.tanima, Shalimer, Shmitra, Siddhartha Ghai, Skier Dude, Skinsmoke, Skylark2008, Skylerb, Sluzzelin, Slysplace, Snthakur, Soman, SomeStranger, Soumitrahazra, Souravdefermat, SpacemanSpiff, Sport woman, Sproy, Spundun, Src2206, Srich32977, Srikeit, Srini81, Stepheng3, Storkk, Subhashischakraborty, Sudipta.kundu, Sukanya92, Sumanch, Sundar, Sunny Gill265, Super cyclist, Supray, Supten, Surajcap, Sushant gupta, Swpb, Taajikhan, Tamilan101, Taratv, Tbhotch, Tbone, Technopilgrim, Template namespace initialisation script, Terissn, Tesi1700, Thaejas, The Magnificent Clean-keeper, The Thing That Should Not Be, Thisthat2011, Thunderboltz, TimBentley, Timberframe, TobyDZ, Tom Edwards, Tom Radulovich, Tony1, Tpbradbury, Trakesht, Travelbird, Treisijs, Trinanjon, Troglo, Tuncrypt, Tutmosis, UnicornTapestry, Unmesh Bangali, Urnonav, UrsusArctosL71, VT hawkeye, Vald, Vale of Glamorgan, Vedran12, Vgranucci, Victor D, Vijayaditya, Wackymacs, Wareq, Wavelength, Whitejay251, Wik, Wiki-uk, Winston365, WoodElf, Woohookitty, World8115, Wouterhagens, Wtmitchell, Wywhpf, Xufanc, Yamamoto Ichiro, Yarnalgo, YellowMonkey, Yorke.aaron, Zakuragi, Zbd, Zeman, Ziaur, Zundark, , ,538 anonymous edits

Gram_panchayat *Source*: http://en.wikipedia.org/w/index.php?title=Gram_panchayat *Contributors*: AdjustShift, Ageo020, Alpha Quadrant (alt), AndrewRT, Andycjp, Angers roams, Apokrif, Auntof6, Avoided, Beyond My Ken, Bittupanpaliya, Bushcarrot, CalJW, Chrisminter, Cybercobra, DBigXray, Dbachmann, Dev, DhruvDhamani, Faradayplank, Feezo, Fieldday-sunday, GTBacchus, Indu, Jackerhack, James500, Ligulem, Loqugr, Lucio Di Madaura, MKar, Nichalp, Nisrec, Phlyght, Priyatu, Qst, Rajankila, Rama's Arrow, Sailing Stonehenge, Salilb, Scopecreep, Shadowjams, Shyamsunder, Spartiate, Spencer, Sskoslia786, Starrahul, Tabletop, ThomasK, Tommy2010, Uncle G, Utcursch, Vprajkumar, WarFox, 87 anonymous edits

Community_development_block_in_India *Source*: http://en.wikipedia.org/w/index.php?title=Community_development_block_in_India *Contributors*: Ageo020, Ansumang, Beyond My Ken, Bogdan Nagachop, Chandan Guha, Explicit, JDP90,

Bhangri_Pratham_Khanda *Source*: http://en.wikipedia.org/w/index.php?title=Bhangri_Pratham_Khanda *Contributors*: Amartyabag, Bijaypatni12, Chandan Guha, GDibyendu, P.K.Niyogi, 1 anonymous edits

Scheduled_castes_and_scheduled_tribes *Source*: http://en.wikipedia.org/w/index.php?title=Scheduled_castes_and_scheduled_tribes *Contributors*: Agsrivaths, Alai, Allen4names, Almithra, Altenmann, Ambuj.Saxena, Andrew Gwilliam, Andycjp, Anwar saadat, Arjun024, Ashokkw, Atul903, Austria156, Babbage, Bigbrothersorder, Borfee, BrokenSegue, CJLL Wright, CarTick, Cdamama, Chancemill, ChrisHodgesUK, Clarkpoon, Cpeel, Dbachmann, Dlempa, Doc Tropics, Dr.K., Dragon guy, Drmies, Dwaipayanc, Earthlyreason, Editor2020, Ekabhishek, Emilio Juanatey, FolkTraditionalist, George Burgess, GeorgeLouis, Gotipe, Hkelkar, Humanrights2011, Humboldt, Idleguy, Indianpeacock, Int21h, J.delancy, Jhartmann, Johari Shauka, JohnLobster, Kaihsu, Kanatonian, Kpalion, LeeHunter, Lhsrhsbvs, Liftarn, Logrithm, Mandarax, Mark Arsten, MatthewVanitas, Mattisse, Mel Etitis, Michael Hardy, Mkrestin, Network for social accountability, Nichalp, Nnemo, Nriweb, Piotrus, Pol430, Prashantroy, Prodiptodas123, Protozoan, QuartierLatin1968, Randhirreddy, Reason turns rancid, Samathawiki, Satendrav80, Sdeepak scor, Shaeed baba Sangat singh ji, SimonP, Sitush, Snowolf, Steinrog2, Tainter, Teleutomyrmex, Truthseeker81, Unnikn, UplinkAnsh, Utcursch, Valentinejoesmith, Valrama, Vinodmikkili, Vkrohit rohit, Wasbeer, Will Beback, Woohookitty, Ybbor, Yogendra Uikey, 167 anonymous edits

Delimitation_Commission_of_India *Source*: http://en.wikipedia.org/w/index.php?title=Delimitation_Commission_of_India *Contributors*: Amarrg, Everyking, Jonoikobangali, Jovianeye, Joy1963, Natrajdr, Neutrality, Nightkey, Rajaramraok, Sadads, Shivap, Shyamsunder, Sting au, The Sage of Stamford, Vinayrajiv, 4 anonymous edits

Dinhata_I_(community_development_block) *Source*: http://en.wikipedia.org/w/index.php?title=Dinhata_I_%28community_development_block%29 *Contributors*: Chandan Guha, Chris the speller

Boundary_delimitation *Source*: http://en.wikipedia.org/w/index.php?title=Boundary_delimitation *Contributors*: BilCat, Drbreznjev, Eastlaw, Firsfron, Good Olfactory, Ground Zero, Incnis Mrsi, Iohannes Animosus, Jnestorius, John Vandenberg, Levineps, Night w, R'n'B, Sadads, Tenmei, 6 anonymous edits

Cooch_Behar_(Lok_Sabha_constituency) *Source*: http://en.wikipedia.org/w/index.php?title=Cooch_Behar_%28Lok_Sabha_constituency%29 *Contributors*: Chandan Guha, GDibyendu, Jonoikobangali, Joy1963, Logicwiki, Sadads, WikiDan61

Image Sources, Licenses and Contributors